NEW HAMPSHIRE
TOTAL ECLIPSE GUIDE

Official Commemorative 2024 Keepsake Guidebook

2024 Total Eclipse State Guide Series

Aaron Linsdau

SASTRUGI PRESS

JACKSON HOLE

Sastrugi Press / Published by arrangement with the author

New Hampshire Total Eclipse Guide: Official Commemorative 2024 Keepsake Guidebook

The author has made every effort to accurately describe the locations contained in this work. Travel to some locations in this book is hazardous. The publisher has no control over and does not assume any responsibility for author or third-party websites or their content describing these locations, how to travel there, nor how to do it safely. Refer to local regulations and laws.

Any person exploring these locations is personally responsible for checking local conditions prior to departure. You are responsible for your own actions and decisions. The information contained in this work is based solely on the author's research at the time of publication and may not be accurate in the future. Neither the publisher nor the author assumes any liability for anyone climbing, exploring, visiting, or traveling to the locations described in this work. Climbing is dangerous by its nature. Any person engaging in mountain climbing is responsible for learning the proper techniques. The reader assumes all risks and accepts full responsibility for injuries, including death.

Sastrugi Press
PO Box 1297, Jackson, WY 83001, United States
www.sastrugipress.com
Quantity sales: Special discounts are available on quantity purchases by corporations, associations, and others. For details, contact the publisher at the address above.

Library of Congress Catalog-in-Publication Data
Library of Congress Control Number: 2019949782
Linsdau, Aaron
New Hampshire Total Eclipse Guide / Aaron Linsdau-1st United States edition
p. cm.
1. Nature 2. Astronomy 3. Travel 4. Photography
Summary: Learn everything you need to know about viewing, experiencing, and photographing the total eclipse in New Hampshire on April 8, 2024.

ISBN-13: 978-1-64922-321-0 (paperback)

508.4—dc23

All photography, maps and artwork by the author, except as noted.
00064
10 9 8 7 6 5

Contents

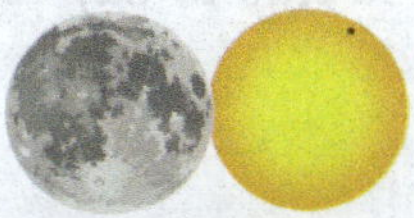

Introduction

Thank you for purchasing this book. It has everything you need to know about the total eclipse in New Hampshire on April 8, 2024.

A total eclipse passing through the United States is a rare event. The last US total eclipse was in 2017. It traveled from Oregon to South Carolina. The last American total eclipse prior to that was in 1979!

The next total eclipse over the US will not be until April 8, 2024. It will pass over Texas, the Midwest, and on to Maine. After that, the next coast-to-coast total eclipse will be in 2045.

It's imperative to make travel plans early. You will be amazed at the number of people swarming to the total eclipse path. Some might say watching a partial versus a total eclipse is a similar experience. It's not.

This book is written for New Hampshire visitors and anyone else viewing the eclipse. You will find general planning, viewing, and photography information inside. Should you travel to the eclipse path in New Hampshire in April, be prepared for an epic trip. The estimates based on the 2017 eclipse suggest that thousands will converge on New Hampshire.

Some hotels in the communities and cities along the path of totality in New Hampshire have already been contacted by people to make reservations. Finding lodging along the eclipse path may be a challenge.

Resources will be stretched far beyond the normal limits. Think gas lines from the late 1970s. It may be likely that traffic along highways will come to a complete standstill during this event. Be prepared with backup supplies.

Many smaller New Hampshire towns are far from any major city. New Hampshire country roads can be slow. Please obey posted speed limits for the safety of everyone. Be cautious about believing a map application's estimate of travel time in New Hampshire.

People in communities along the path of the total eclipse may rent out properties for this event. With this major celestial spectacle in the spring of 2024, be assured that New Hampshire "hasn't seen anything yet."

Is this to say to avoid New Hampshire or other areas during the eclipse? Not at all! This guidebook provides ideas for interesting, alternative, and memorable locations to see the eclipse. It will be too late to rush to a better spot once the eclipse begins. Law enforcement will be out to help drivers reconsider speeding.

Please be patient and careful. There will be a large rush of people from all over the world, converging on New Hampshire to enjoy the total eclipse. Be mindful of other drivers on eclipse weekend, as they may not be familiar with New Hampshire roads.

You should feel compelled to play hooky on April 8. Ask for the day off. Take your kids out of school. They'll be adults before the next chance to see a total eclipse over America. Create family memories that will last a lifetime. Sastrugi Press does not normally advocate skipping school or work. Make an exception because this is too big an event to miss.

Wherever you plan to be along the total eclipse path, leave early and remember your eclipse glasses. People from all around the planet will converge on New Hampshire. Be good to your fellow humans and be safe. We all want to enjoy this spectacular show.

Visit www.sastrugipress.com/eclipse for the latest updates for this state eclipse book series.

Author Information

Polar explorer and motivational speaker Aaron Linsdau's first book, *Antarctic Tears,* is an emotional journey into the heart of Antarctica. He ate two sticks of butter every day to survive. Aaron coughed up blood early in the expedition and struggled with equipment failures. Despite the endless difficulties, he set a world record for surviving the longest solo expedition to the South Pole.

Aaron teaches how to build resilience to overcome adversity by managing attitude. He shares his techniques for overcoming adrenaline burnout and constant overload. He inspires audiences to face their challenges with a new perspective. As a motivational speaker, Aaron talks about courage, resilience, attitude, safety, and risk. He hopes that you will be inspired and have an enjoyable time watching the total eclipse in New Hampshire. Visit www.aaronlinsdau.com to learn more.

All About New Hampshire

Overview of New Hampshire

New Hampshire, known as the Granite State, is riddled with granite quarries. But, that's not all it has to offer. Filled with quaint little towns, the White Mountains, and even a few theme parks, New Hampshire is a wonderful place to visit. It's perfect for a weekend getaway, but there is so much to see and do you'll want to stay much longer.

New Hampshire is known for its wilderness and outdoor activities. There's no shortage in New Hampshire of places to camp, hike, bike, ski, or snowboard.

Lake Winnipesaukee is located in central New Hampshire and is a great place for you and your family to visit. The Lake is seventy-two square miles, and has over two hundred sixty islands within the lake!

Northern New Hampshire is filled with campgrounds, mountains

to hike in the summer, and skiing in the winter. When hiking, you're sure to encounter wildlife like birds, moose, black bears, and deer. In the fall, you'll be sure to see the gorgeous fall foliage with any hike, bike, or car ride throughout the state.

The White Mountains are located in northern New Hampshire, where you'll find Mount Washington, the tallest mountain in the northeast. The Appalachian Trail crosses through the summit of Mount Washington. Mount Washington is known for its crazy weather at its summit, where the Mount Washington Observatory is located. You can choose to hike Mount Washington or use the auto road to climb to the summit. You can also use the Mount Washington Cog Railway to reach the summit. There are bike races and runs on the mountain and skiing in the winter.

Located near Mount Washington is the town of Jefferson, New Hampshire. Here you'll find a quaint little town with a large golf course, plenty of bed-and-breakfasts and campgrounds, and perhaps the best part—Santa's Village. Santa's Village is a small Christmas-themed theme park with rides for kids of all ages, a small water park, great food, and of course, Santa and Mrs. Claus! Christmas music is played throughout the park, performances in the Polar Theater, and a workshop where kids can make their own souvenirs.

Not too far away is Glenn, New Hampshire, where you'll find another theme park for kids, Story Land. While Story Land's theme is nursery rhymes and fairy tales and seems like it would be fun just for little ones, they have attractions and rides for the whole family. They offer great food, amazing shows, water rides, and of course they have all your favorite characters! Story Land offers sensory-friendly dates throughout the year as well. Story Land is large, and you'll need more than one day to enjoy all the attractions.

In nearby North Conway you'll find the Conway Scenic Railroad. There are multiple trips to choose from, ranging from just an hour long to more than five hours long. There is dining available on the trips, and they hold special events throughout the year for all ages. They include murder mystery dinner rides, fall-themed rides, and in December, they have rides to the North Pole and train rides with Santa.

In Salem, New Hampshire, you'll find Canobie Lake Park, a theme park geared more toward adults and teenagers. They do offer activities for young ones, but older kids will have the best time here. Water rides, great food, amazing roller coasters, and an arcade are some of what you'll find.

If you like shopping, New Hampshire has you covered. First there's no sales tax. And with multiple outlet shopping malls in the state, you'll be sure to find great deals!

The small towns in New Hampshire have some intriguing local stores filled with locally made items from candy to candles. You'll also find some tasty local eateries in these small towns.

There's never enough time to explore the state of New Hampshire.

There are so many places to see and things to do. If you're visiting for the eclipse in 2024, you'll want to plan your trip to include some of these amazing places. Also, be sure to include some time for just driving through the state and exploring on your own. You never know what you'll discover.

Hotels and Motels During the Eclipse

Once excitement of the total eclipse over New Hampshire spreads, rooms will become scarce. Many hotels in towns along the path of totality in western states sold out for a year or more during the 2017 total eclipse. New Hampshire is not alone in this challenge. Hotels all along the path of totality will sell out in anticipation of the 2024 total eclipse.

What does this mean for eclipse visitors? Lodging and room rentals in eclipse towns will be at a massive premium. Does that mean all hope is lost to find a place to stay? Not at all. But you will have to be creative. There will be few, if any, hotel rooms available in these eclipse cities by early 2024. Accommodations in the cities and towns along the path of the eclipse will be difficult to come by.

In summer 2017, the author searched on Hotels.com for rooms along the 2017 total eclipse path on the weekend of August 21 and found many major cities sold out. Once word of the 2024 eclipse spreads, room rates will increase and availability will drop.

Search for rooms farther away from the eclipse path. If you are willing to stay in cities outside the eclipse path, you will have better success at finding rooms. As the eclipse approaches, people will book rooms farther from the totality path. By early spring, rooms in cities near the total eclipse path may be unavailable. The effect of this event will be felt across New Hampshire and the rest of the United States.

Think regionally when looking for rooms. Be prepared to search far and wide during this major event. If a five-hour drive is manageable, your lodging options greatly expand, but it also increases your travel risk.

Internet Rentals

To find rooms to stay in towns along the eclipse path, try a web service such as Airbnb.com. Note that some people rent out rooms or homes illegally, against zoning regulations. Cities will feel the crunch of inquiries early due to others who experienced the 2017 eclipse.

If cities fully enforce zoning laws, authorities may prevent your weekend home rental. Online home rentals during the eclipse will be a target for rental scams. People from out of the area steal photos and descriptions, then post the home for rent. You send your check or wire money to a "rental agent" then show up to find you have been scammed. If the deal sounds strange or too good to be true, run away.

Camping

If you can book a campsite, do it as soon as you can. Do not wait. All areas in the national forests are first-come, first-served. Forest roads may be packed. Expect all areas to be swarming with people. Show up early to stake out your spot. Consider staying farther away and driving early on April 8.

Please respect private land too. New Hampshire folks don't take kindly to people overrunning their property without permission. In a big state with millions of residents, people are very protective, but they're friendly, too. You never know what you might be able to arrange with a smile and a bit of money.

This all said, there are plenty of camping opportunities throughout New Hampshire. You don't have to sleep exactly on the eclipse path. If you're ready to rough it, there are national forest camping options.

Government agencies will meet years in advance to talk about how to manage the influx of people. Every possible government agency will be working full time to enforce the various rules and regulations.

National Parks and Monuments

Finding a camping site at any state park, national park, or national monument along the eclipse path in New Hampshire will be challenging. To watch the eclipse from any location, you do not have to sleep in it. You just need to drive there in the morning.

Law enforcement will be present on the eclipse weekend. Hundreds of thousands of people are expected in the region. Parking may overflow. It will make parking lots and lines on Black Friday at the mall look uncrowded. For an event of this magnitude, find your location as early as possible.

The first sentence of the national parks mission statement is:

> *"The National Park Service preserves unimpaired the natural and cultural resources and values of the national park system for the enjoyment, education, and inspiration of this and future generations."*

Roadside camping (sleeping in your car) is not allowed in national monuments or parks. Park facilities are only designed to handle so many people per day. Water, trash collection, and toilets can only withstand so much. If you notice trash on the ground, take a moment to throw it away. Protect your national park and help out. Rangers are diligent and hardworking but they can only do so much to manage the expected crowds.

National and State Forests

There are national and state forest options in New Hampshire. They all have camping opportunities. The forest service manages undeveloped and primitive campsites. Be sure to check for any fire restrictions. Check with individual agencies for last-minute information and regulations. The forest service requires proper food storage. Plan to purchase food and water before choosing your campsite. Below is a partial list of forests and parks in New Hampshire:

White Mountain National Forest:
https://www.fs.usda.gov/whitemountain
Fox State Forest
Gile State Forest
Low State Forest

Backcountry service roads abound in New Hampshire. Maps for forests are available at local visitor centers and bookstores. This book's website has digital copies of some forest maps.

Printed national forest maps are large and detailed. They have illustrated road paths, connections, and other vital travel information not available on digital device maps. Viewing digital maps on your smartphone or mobile pad is difficult. If you plan to camp in the forest, a real paper map is a wise investment.

Camping in federal wilderness areas is also allowed. Those areas afford the ultimate backcountry experience. However, be aware that no vehicle travel is allowed in the specially designated areas. This ban includes: vehicles, bikes, hang gliders, and drones. You can travel only on foot or with pack animals.

Sleep in Your Car

Countless RVs, campers, trucks, cars, and motorcycles will flood New Hampshire. Sleeping in your car with friends is tolerable. Doing so with unadventurous spouses or children is another matter.

Do not be caught along the path of the total eclipse without some sort of plan, especially in the bigger cities of New Hampshire. The whole path of totality will fill with people on April 8.

Useful Local Webcams

Local webcams are handy to make last-minute travel decisions. Modern webcams are sensitive enough to show headlights at night. Use them to determine if there are issues before traveling out. Eclipse traffic will add to the morning commuter traffic.

There are smartphone applications which are useful to check webcams in many locations. Consult your device's app store for the latest updates. Whether you use an app or computer, an Internet search will reveal many handy webcams for your eclipse planning.

Weather

It's all about the weather during the eclipse. Nothing else will matter if the sky is cloudy. You can be nearly anywhere along the path in New Hampshire and catch a view of the event when traffic comes to a standstill. But if there's a cloud cover forecast, seriously reconsider your viewing location.

Travel early wherever you plan to go. Attempting to change locations an hour before the eclipse due to weather will likely cause you to miss the event. New Hampshire country roads can be narrow and slow. The number of vehicles will cause unexpected backups.

Modern Forecasts

Use a smartphone application to check the up-to-date weather. Wunderground is a good application and has relatively reliable forecasts for the region. The hourly forecast for the same day has been rather accurate for the last two years. The below discussion refers to features found in the Wunderground app. However, any application with detailed weather views will improve your eclipse forecasting skills.

Cloud Cover Forecast

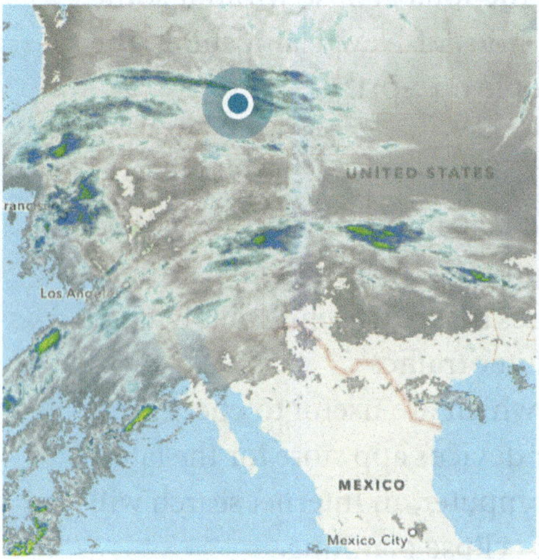

Infrared cloud map showing the worst case eclipse cloud cover. Courtesy of National Weather Service.

The most useful forecast view is the visible and infrared cloud-coverage map. Avoid downloading this app the night before and trying to learn how to read it. Practice reading them at home. It's imperative to understand how to interpret the maps early.

All cloud cover, night or day, will appear on an infrared map. Warm, low-altitude clouds are shown in white and gray. High-altitude cold clouds are displayed in shades of green, yellow, red, and purple. Anything other than a clear map spells eclipse-viewing problems.

To improve your weather guess, use the animated viewer of the cloud cover. It will give you a sense of cloud motion. You can discern whether clouds or rain are moving toward, away from, or circulating around your location.

Normal New Hampshire Weather Pattern

Due to the direction of the jet stream, most weather travels across the Pacific Ocean, through the western states, over the Great Lakes, and then into New Hampshire. On occasion, weather can approach from Canada or the Atlantic. Due to the nature of the storms from the Arctic, weather in New Hampshire can be unpredictable in the spring.

The common weather pattern in April is slightly warm in the afternoon and cool to cold in the evenings. Passing cold fronts in spring can bring regular cloud cover and rain showers.

Historically, New Hampshire tends to have moderate cloud cover during April. Prepare to make adjustments. If anything other than clear skies are predicted, drive to other parts of New Hampshire, Vermont, or Maine.

Be aware of severe weather in New Hampshire. Although spring is well underway by the time of the total eclipse, there have been many substantial storms in April. Pay attention to the weather forecast. If dangerous weather is predicted, your main concern should be safety rather than chasing an eclipse.

Consider that slow-moving clouds can obscure the sun for far longer than the local duration of the totality. The time of totality is so short that you do not want to risk it. Missing it due to a single cloud will be a major disappointment.

Local Eclipse Weather Forecasts

Local town and city newspapers, radio, and television stations around New Hampshire will have a weekend edition with articles discussing the eclipse weather. However, conditions change unpredictably in New Hampshire. A three-day forecast for April may be incorrect.

Finding the Right Location to View Eclipse Effects

One of the peculiarities of total eclipses is that the entire show is not only in the sky. There are other unusual effects seen during the total eclipse that are worth looking for.

The first effect to watch for is the crescent moon shapes created from leaf shadows on the ground. They're best viewed on a sidewalk or asphalt. They can only be observed during the partial eclipse. The other effect that is worth watching for is the shadow bands or "snakes" as they're commonly called.

Shadow banding is seen right before and after the totality takes place. They're best observed on smooth, plain-colored surfaces. If you plan to be in the forest for the eclipse, you may struggle to see the bands but will likely see crescent shadows all around on the ground.

One of the supreme challenges with all of these effects is choosing what to watch. You can see the crescent shadows in the hour before and after the totality but shadow banding happens before or after totality. It is more difficult to look away from the eclipse than you think.

Road Closures and Traffic

Highways connecting various New Hampshire towns in the total eclipse path will be heavily impacted on the weekend before and day of the eclipse. As was found in the 2017 total eclipse, there is no way to predict which areas will be impacted.

Planning ahead is essential to give you the best opportunity to enjoy the eclipse without the nightmare of being stuck in traffic for hours on end. The traffic in states during the 2017 eclipse was stunning, so it may be substantial in New Hampshire in 2024.

Update yourself with the latest road report information from the New Hampshire road condition website:

https://newengland511.org/

It's imperative to plan early and have one if not more backup plans in case of difficult travel conditions. April weather is unpredictable and variable.

If you believe it's necessary to leave a town to watch the eclipse, do so the night before or extremely early in the morning. RVs are common, and trains of them crawl through popular areas.

Communication Information

Cellular Phones

Cellular "cell" phone service in some remote New Hampshire locations may be problematic. Most of the time there is good coverage along the main highways and interstates. However, even along major thoroughfares, there can be little or no coverage.

It's possible to find zones where text messages will send when phone calls are impossible. If you cannot make a phone call, the chance of having data coverage for web surfing or e-mail is low.

Please look up any information or communicate what you need before departing from the main roads around New Hampshire. Bureau of Land Management (BLM) areas sometimes have coverage. Planned to be self-contained. Plan for your cell phone not to connect.

You may find yourself out of cell service. With a large number of cell users in a concentrated area, coverage and data speed may collapse as well. Search on the phrase "cell phone coverage breathing".

Wilderness and Forest Safety

All New Hampshire forest and wilderness areas are full of wild animals. Although beautiful, wild animals can be dangerous. They can easily injure or kill people, as they are far more powerful than humans. Do not try to feed any wild animals, including squirrels, foxes, and chipmunks, as they can carry diseases. These suggestions apply to all public lands.

Ticks

Ticks exist all across the United States, but not all species transmit disease. Ticks cannot fly or jump, but they climb grasses in shrubs in order to attach to people or animals that pass by. Ticks feed on the blood of their host. In doing so, they can transmit potentially life-threatening diseases such as Lyme disease.

Spiders

Although the chance of encountering a venomous spider is low, it is not uncommon to encounter them. The two dangerous spiders in the state are the black widow and brown recluse. Should you encounter either species, simply leave it alone. If you are bitten, seek immediate medical attention, as their toxin can be life-threatening.

Venomous Snakes

There is one known species of venomous snake in New Hampshire, the Timber Rattlesnake. Although these reptiles are not generally aggressive and rarely seen, they can strike when provoked or threatened. Of the approximately 8,000 people annually bitten by venomous snakes in the United States, ten to fifteen people die according to the U.S. Food and Drug Administration.

The best way to avoid rattlesnake encounters is to be mindful of your environment. Do not place your hands or feet in locations where you cannot clearly see the surroundings. Avoid heavy brush or tall weeds where snakes hide during the day. Step on a log or rock rather than over it, as a hidden snake might be on the other side. Rattlesnakes may not make any noise before striking.

Avoid handling all snakes. Should you be bitten, stay calm and call 911 or emergency dispatch as soon as possible. Transport the victim to the nearest medical facility immediately. Rapid professional treatment is the best way to manage rattlesnake bites. Refer to US Forest Service and professional medical texts for more information on managing rattlesnakes injuries.

Bears

The forests of New Hampshire are potentially home to black bears. Though they are listed as not often seen, they have been sighted in the state. Although they often appear docile, they can become aggressive if threatened. In the unlikely event of an attack, fight back against the bear. Use whatever you have at your disposal to defend yourself. Report all negative or aggressive bears to the local authorities.

If a bear hears you, it will usually vacate the area. Bear charges are often caused by unexpected and surprise encounters. Noise is the best defense to avoid surprising bears. Regularly clap, make noise, and talk loudly. New Hampshire Fish & Game is a good starting point to learn more about black bears at https://www.wildlife.state.nh.us/wildlife/profiles/black-bear.html.

It is recommended to stay one hundred yards (300 feet) away from all bears. They are exciting to see but need their space. Refer to current forest or park regulations for more safety information.

Mountain Lions

Though declared to be extinct in the state, there have been reports of mountain lions by various locals and news agencies. If you encounter a mountain lion, do not run. Keep calm, back away slowly, and maintain eye contact. Do all you can to appear larger. Stand upright, raise your arms, or hoist your jacket. Never bend over or crouch down. If attacked, fight back.

Eclipse Day Safety

1. Hydrate

Spring temperatures are usually mild to warm. The excitement of the event can distract you from managing hydration. Drink plenty of water. Consume more than you would at home.

2. Eye Safety time

Use certified eclipse safety glasses at all times when viewing the partial eclipse. Only remove the glasses when the totality happens. Give your eyes time to rest. They can dry out and become irritated. Bring FDA approved eye drops to keep your eyes moist.

3. Sun exposure

Facing at the sun for three hours can result in sunburns. Wear sunglasses and liberally apply sunscreen to avoid sunburns.

4. Eat well

Keep your energy up. Appetite loss is common when traveling. Maintain your normal eating schedule.

5. Prepare for temperature changes

Temperatures will drop rapidly during the eclipse and also once the sun sets. Bring appropriate clothing.

6. Talk with your doctor

If the humidity or heat bothers you talk with your doctor before traveling. Seek professional medical attention for serious symptoms.

Total eclipse path across the United States (approximate).

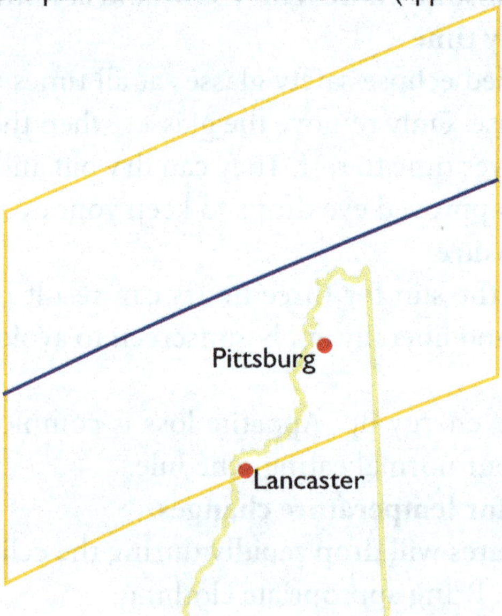

Total eclipse path across New Hampshire (approximate).

All About Eclipses

How an Eclipse Happens

An eclipse occurs when one celestial body falls in line with another, thus obscuring the sun from view. This occurs much more often than you'd think, considering how many bodies there are in the solar system. For instance, there are over 150 moons in the solar system. On Earth, we have two primary celestial bodies: the sun and the moon. The entire solar system is constantly in motion, with planets orbiting the sun and moons orbiting the planets. These celestial bodies often come into alignment. When these alignments cause the sun to be blocked, it is called an eclipse.

For an eclipse to occur, the sun, Earth, and moon must be in alignment. There are two types of eclipses: solar and lunar. A solar eclipse occurs when the moon obscures the sun. A lunar eclipse occurs when the moon passes through Earth's shadow. Solar eclipses are much more common, as we experience an average of 240 solar eclipses a century compared to an average of 150 lunar eclipses. Despite this, we are more likely to see a lunar eclipse than a solar eclipse. This is due to the visibility of each.

For a solar eclipse to be visible, you have to be in the moon's shadow. The problem with viewing a total eclipse is that the moon casts a small shadow over the world at any given time. You have to be in

* ILLUSTRATION NOT TO SCALE

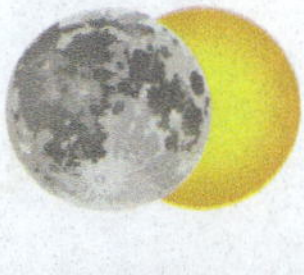

a precise location to view a total eclipse. The issue that arises is that most of these locations are inaccessible to most people. Though many would like to see a total solar eclipse, most aren't about to set sail for the middle of the Pacific Ocean. In fact, a solar eclipse is visible in the same place on the world on average every 375 years. This means that if you miss a solar eclipse above your hometown, you're not going to see another one unless you travel or move.

It's much easier to catch a glimpse of a lunar eclipse, even though they occur at a much lower frequency than their solar counterparts. A lunar eclipse darkens the moon for a few hours. This is different than a new moon when it faces away from the sun. During these eclipses, the moon fades and becomes nearly invisible.

Another result of a lunar eclipse is a blood moon. Earth's atmosphere bends a small amount of sunlight onto the moon turning it orange-red. The blood moon is caused by the dawn or dusk light being refracted onto the moon during an eclipse.

Lunar eclipses are much easier to see. Even when the moon is in the shadow of Earth, it's still visible throughout the world because of how much smaller it is than Earth.

Total vs. Partial Eclipse

What is the difference between a partial and total eclipse? A total eclipse of either the sun or the moon will occur only when the sun, Earth, and the moon are aligned in a perfectly straight line. This ensures that either the sun or the moon is partially or completely obscured.

In contrast, a partial eclipse occurs when the alignment of the three celestial bodies is not in a perfectly straight line. These types of eclipses usually result in only a part of either the sun or the moon being obscured. This is often what led to ancient civilizations believing that some form of magical beast or deity was eating the sun or the moon. It appears as though something has taken a bite out of either the sun or the moon during a partial eclipse.

Total eclipses, rarer than partial eclipses, still occur quite often. It's more difficult for people to be in a position to experience such an event firsthand. Total solar eclipses can only be viewed from a small portion of the world that falls into the darkest part of the moon's shadow. Often this happens in the middle of the ocean.

The Moon's Shadow

The moon's shadow is divided into two parts: the umbra and the penumbra. The former is much smaller than the latter, as the umbra is the innermost and darkest part of the shadow. The umbra is thus the central point of the moon's shadow, meaning that it is extremely small in comparison to the entire shadow. For a total solar eclipse to be visible, you need to be directly beneath the umbra of the moon's shadow. This is because that is the only point at which the moon completely blocks the view of the sun.

In contrast, the penumbra is the region of the moon's shadow in which only a portion of the light cast by the sun is obscured. When

Total eclipse shadow 2016 as seen from 1 million miles on the Deep Space Climate Observatory satellite. Courtesy of NASA.

standing in the penumbra, you are viewing the eclipse at an angle. In the penumbra, the moon does not completely block the sun from view. This means that while the event is a total solar eclipse, you'll only see a partial eclipse. The umbra for the April 8 eclipse is over one hundred miles wide. The penumbra will cover much of the United States.

To provide some context, one total solar eclipse we experienced occurred on March 9, 2016, and was visible as a partial eclipse across most of the Pacific Ocean, parts of Asia, and Australia. However, the only place in the world to view this total solar eclipse was in a few parts of Indonesia.

Due to the varied locations and the brief periods for which they're visible, it's difficult to see each and every eclipse that occurs. The umbra of the moon is such a small fraction of the entire shadow and the majority of our planet is comprised of water. Thus, the rarity of being able to view a total solar eclipse increases significantly because it's likely that the umbra will fall over some part of the ocean rather than a populated landmass. There are not total or annular eclipses every month because the moon's orbit is 5.1° off the ecliptic plane of the Earth and sun.

Eclipses Throughout History

Ancient peoples believed eclipses were from the wrath of angry gods, portents of doom and misfortune, or wars between celestial beings. Eclipses have played many roles in cultures, creating myths since the dawn of time. Both solar and lunar eclipses affected societies worldwide. Inspiring fear, curiosity, and the creation of legends, eclipses have cast a long shadow in the collective unconscious of humanity throughout history.

Early Myth & Astronomy

Documented observations of solar eclipses have been found as far back in history as ancient Egyptian and Chinese records. Timekeeping was important to ancient Chinese cultures. Astronomical observations were an integral factor in the Chinese calendar. The first observation of a solar eclipse is found in Chinese records from over 4,000 years ago. Evidence suggests that ancient Egyptian observations may predate those archaic writings.

Many ancient societies, including Roman, Greek and Chinese civilizations, were able to infer and foresee solar eclipses from astronomical data. The sudden and unpredictable nature of solar eclipses had a stressful and intimidating effect on many societies that lacked the scientific insight to accurately predict astronomical events. Relying on the sun for their agricultural livelihood, those societies interpreted solar eclipses as world-threatening disasters.

In ancient Vietnam, solar eclipses were explained as a giant frog eating the sun. The peasantry of ancient Greece believed that an eclipse was the sign of a furious godhead, presenting an omen of wrathful retribution in the form of natural disasters. Other cultures were less speculative in their investigations. The Chinese Song Dynasty scientist Shen Kuo proved the spherical nature of the Earth and heavenly bodies through scientific insight gained by the study of eclipses.

The Eclipse in Native American Mythology

Eclipses have played a significant role in the history of the United States. Before Europeans settled in the Americas, solar eclipses were important astronomical events to Native American cultures. In most native cultures, an eclipse was a particularly bad omen. Both the sun and the moon were regarded as sacred. Viewing an eclipse, or even being outside for the duration of the event, was considered highly taboo by the Navajo culture. During an eclipse, men and women would simply avert their eyes from the sky, acting as though it was not happening.

The Choctaw people had a unique story to explain solar eclipses. Considering the event as the mischievous actions of a black squirrel and its attempt to eat the sun, the Choctaw people would do their best to scare away the cosmic squirrel by making as much noise as possible until the end of the event, at which point cognitive bias would cause them to believe they'd once again averted disaster on an interplanetary scale.

Contemporary American Solar Phenomena

The investigation of solar phenomena in twentieth-century American history had a similarly profound effect on the people of the United States. A total solar eclipse occurring on the sixteenth of June, 1806, engulfed the entire country. It started near modern-day Arizona. It passed across the Midwest, over Ohio, Pennsylvania, New York, Massachusetts, and Connecticut. The 1806 total eclipse was notable for being one of the first publicly advertised solar events. The public was informed beforehand of the astronomical curiosity through a pamphlet written by Andrew Newell entitled *Darkness at Noon, or the Great Solar Eclipse*.

This pamphlet described local circumstances and went into great detail explaining the true nature of the phenomenon, dispelling myth and superstition, and even giving questionable advice on the best methods of viewing the sun during the event. Replete with a short historical record of eclipses through the ages, the *Darkness at Noon* pamphlet is one of the first examples of an attempt to capitalize on the mysterious nature of solar eclipses.

Another notable American solar eclipse occurred on June 8, 1918. Passing over the United States from Washington to Florida, the eclipse was accurately predicted by the U.S. Naval Observatory and heavily documented in the newspapers of the day. Howard Russell Butler, painter and founder of the American Fine Arts Society, painted the eclipse from the U.S. Naval Observatory, immortalizing the event in *The Oregon Eclipse*.

Four more total solar eclipses occurred over the United States in the years 1923, 1925, 1932, and 1954, with another occurring in 1959. The October 2, 1959, solar eclipse began over Boston, Massachusetts. It was a sunrise event that was unviewable from the ground level. Eminent astronomer Jay Pasachoff attributed this event to sparking his interest in the study of astronomy. Studying under Professor Donald Menzel of Williams College, Pasachoff was able to view the event from an airline hired by his professor.

To this day, many myths surround the eclipse. In India, some local customs require fasting. In eastern Africa, eclipses are seen as a danger to pregnant women and young children. Despite the mystery and legend associated with unique and rare astronomical events, eclipses continue to be awe-inspiring. Even in the modern day, eclipses draw out reverential respect for the inexorable passing of celestial bodies. They are a reminder of the intimate relationship between the denizens of Earth and the universe at large.

Present Day Eclipses

The year 2017 brought the world's most-watched total eclipse in history on August 21, when a total solar eclipse crossed the United States. An annular eclipse, a "ring of fire," will pass over the United States in 2023 from Oregon to Texas. Though impressive, it will not

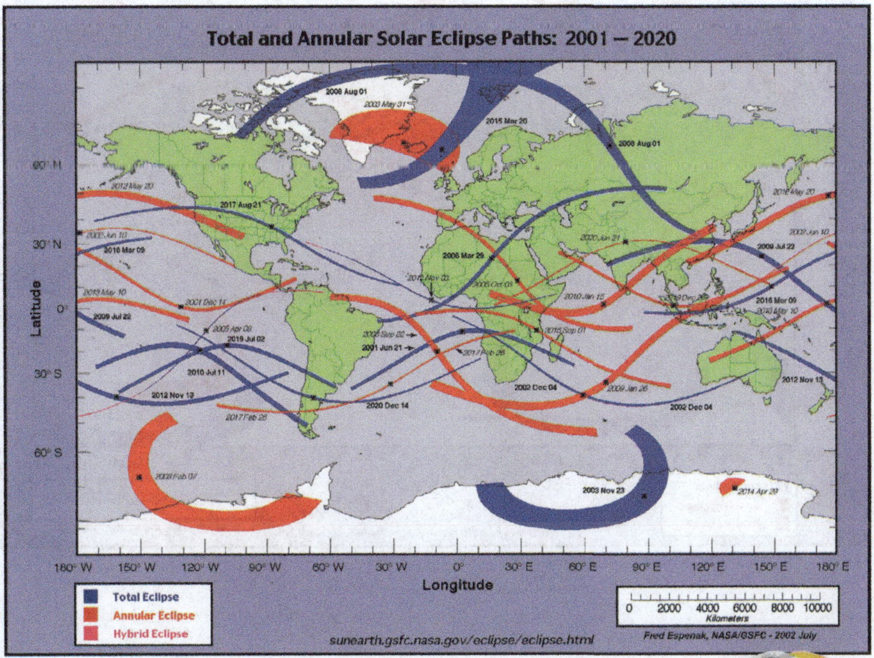

COURTESY OF NASA

compare to the 2024 total eclipse. There is little in nature that equals the spectacle of the sun's corona and seeing stars in the day.

There will be multiple partial, annular, or hybrid eclipses across the world before the 2024 total eclipse. However many are in remote, inaccessible, or potentially dangerous locations on the globe. In 2019 and 2020, Chile and Argentina will experience total eclipses. The next total eclipse after that will occur over Antarctica in 2021. An extremely rare hybrid eclipse will happen in 2023 over the Indian Ocean, Australia, and Indonesia.

The next total solar eclipse viewable from the United States will occur on April 8, 2024. It will be visible in fifteen states: Texas, Oklahoma, Arkansas, Missouri, Tennessee, Kentucky, Illinois, Indiana, Ohio, Pennsylvania, Michigan, New York, Vermont, New Hampshire, and Maine.

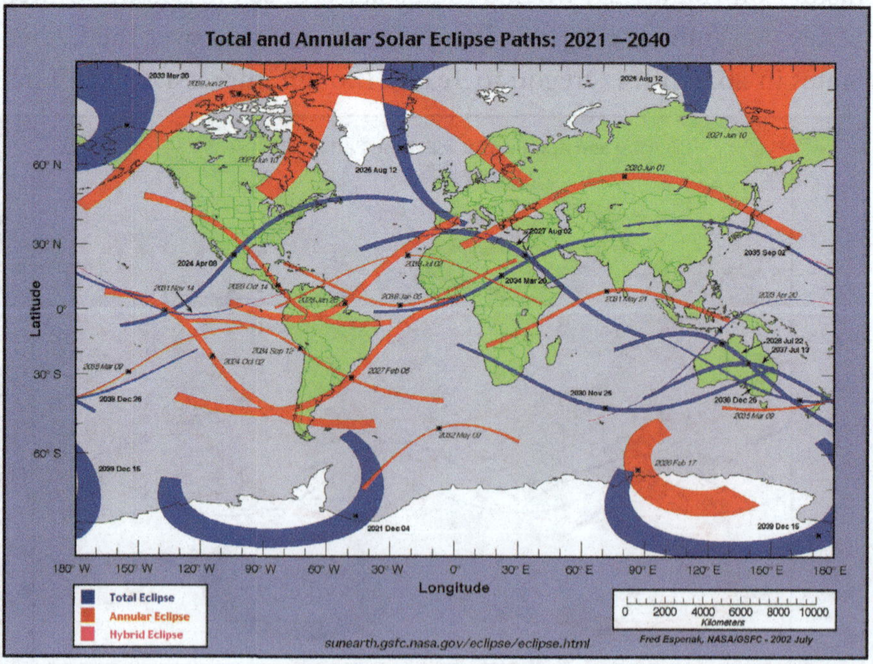

COURTESY OF NASA

Viewing and Photographing the Eclipse

At-home Pinhole Method

Use the pinhole method to view the eclipse safely. It costs little but is the safest technique there is. Take a stiff piece of single-layer cardboard and punch a clean pinhole. Let the sun shine through the pinhole onto another piece of cardboard. That's it!

Never look at the sun through the pinhole. Your back should be toward the sun to protect your eyes. To brighten the image, simply move the back piece of cardboard closer to the pinhole. To see it larger, move the back cardboard farther away. Do not make the pinhole larger. It will only distort the crescent sun.

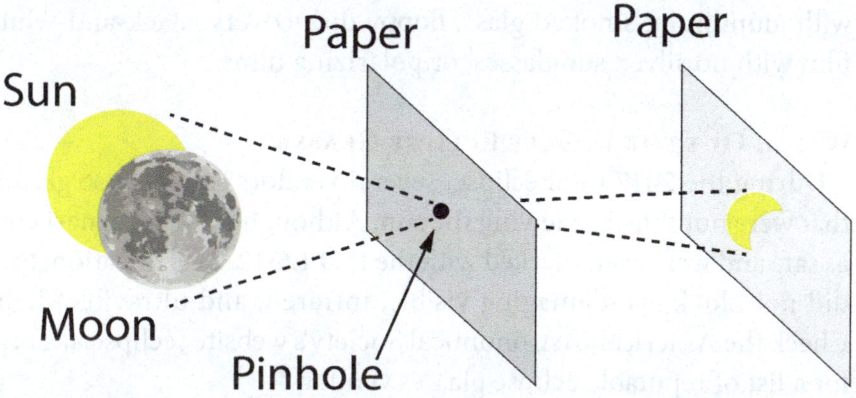

Welding Goggles

Welding goggles that have a rating of fourteen or higher are another useful eclipse viewing tool. The goggles can be used to view the solar eclipse directly. Do not use the goggles to look through binoculars or telescopes, as the goggles could potentially shatter due to intense direct heat. Avoid long periods of gazing with the goggles. Look away every so often. Give your eyes a break.

Solar Filters for Telescopes

The ONLY safe way to view solar eclipses using telescopes or binoculars is to use solar filters. The filters are coated with metal

to diminish the full intensity of the sun. Although the filters can be expensive, it is better to purchase a quality filter rather than an inexpensive one that could shatter or melt from the heat.

The filters attach to the front of the telescope for easy viewing. Remember to give your telescope cooling breaks. Rapid heating can damage your equipment with or without filters attached.

Watch Out for Unsafe Filters

There are several myths surrounding solar filters for eclipse viewing. In order for filters to be safe, they must be specially designed for looking at a solar eclipse. The following are all unsafe for eclipse viewing and can lead to retinal damage: developed colored or chromogenic film, black-and-white negatives such as X-rays, CDs with aluminum, smoked glass, floppy disk covers, black-and-white film with no silver, sunglasses, or polarizing films.

Watch Out for Unsafe Eclipse Glasses

During the 2017 total eclipse, several vendors sold eclipse glasses that were not safe for viewing the sun. Although they were marketed as safe and were even marked with the ISO 12312-2 certification, they did not block eye-damaging visible, infrared, and ultraviolet light. Check the American Astronomical Society's website (eclipse.aas.org) for a list of reputable eclipse glasses vendors.

Viewing with Binoculars

When viewing the eclipse with binoculars, it is important to use solar filters on both lenses until totality. Only then is it safe to remove the filter. As the sun becomes visible after totality, replace the filters for safe viewing. Protect your pupils. Remember to give your binoculars a cool-down break between viewings. They can overheat rapidly from being pointed directly at the sun even with filters attached.

Planning Ahead

There are many things to keep in mind when viewing a total eclipse. It is important to plan ahead to get the most out of this extraordinary experience.

UNDERSTANDING SUN POSITION

All compass bearings in this book are true north. All compasses point to Earth's magnetic north. The difference between these two measurements is called magnetic declination. The magnetic declination for New Hampshire is:

14° 7' W ± 0° 23' (for Concord in 2024)

Adjust the declination from the azimuth bearing as given in the text, and set your compass to that direction.

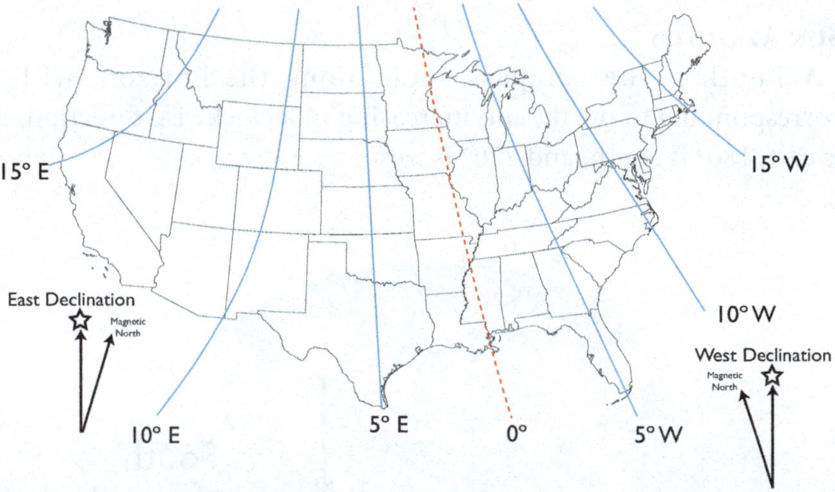

If you purchase a compass with a built-in declination adjustment, you can change the setting once and eliminate the calculations. The Suunto M-3G compass has this correction. A compass with a sighting mirror or wire will help you make a more accurate azimuth sighting.

The Suunto M-3G also has an inclinometer. This allows you to measure the elevation of any object above the horizon. Use this to figure out how high the sun will be above your position.

You can also use a smartphone inclinometer and compass for this purpose. Make sure to calibrate your smartphone's compass before every use, otherwise it might indicate the wrong bearing. Set the smartphone compass for true north to match the book. Understand the compass prior to April 8. There will be little time to guess or search on the Internet. Smartphone and GPS compasses are "sticky." Their compasses don't swing as freely as a magnetic compass does.

The author has used his magnetic compass for azimuth measurements and a smartphone to measure elevation. Combining these two tools will allow you to make the best sightings possible.

Outdoor sporting goods stores in most towns and cities carry compasses. Purchase and practice with a good compass in your hometown well before the event. Take the time to learn how to use it before the day of the eclipse. You do not want to struggle with orienteering basics under pressure.

Sun Azimuth

Azimuth is the compass angle along the horizon, with 0° corresponding to north, and increasing in a clockwise direction. 90° is east, 180° is south, and 270° is west.

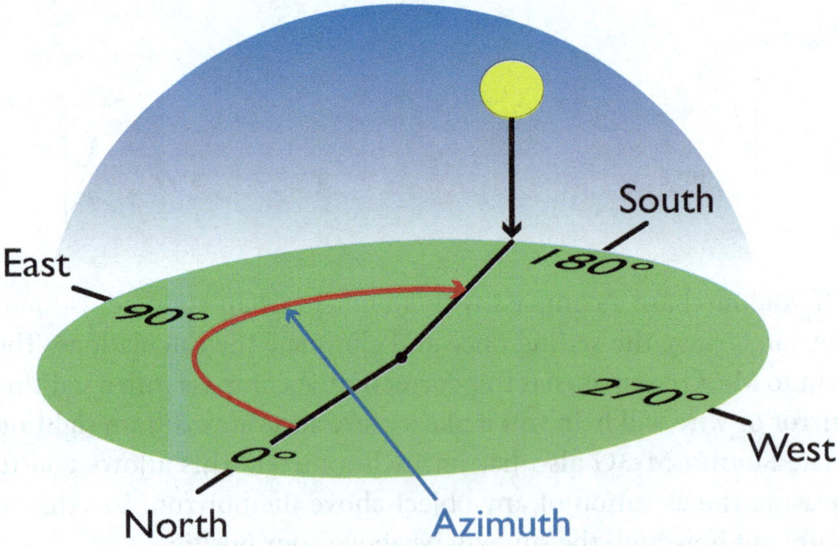

Sun Elevation

Altitude is the sun's angle up from the horizon. A 0° altitude means exactly on the horizon and 90° means "straight up."

Using the sun azimuth and elevation data, you can predict the position of the sun at any given time. Positions given in this book coincide with the time of eclipse totality unless otherwise noted.

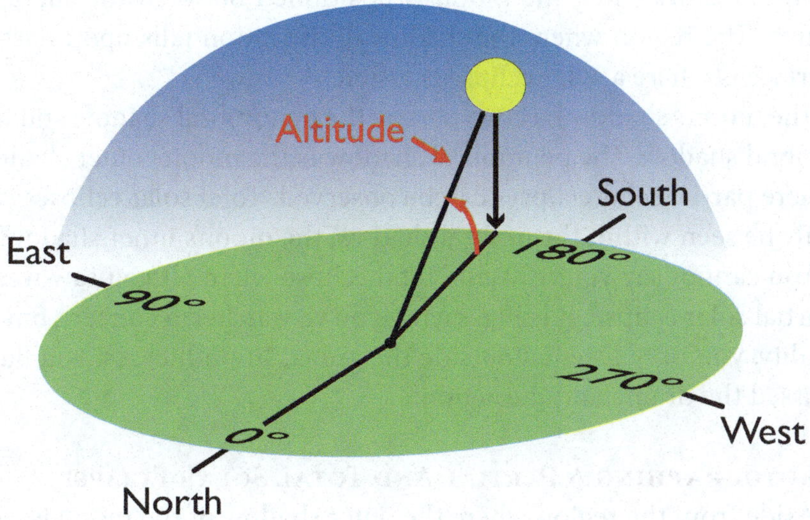

Eclipse Data for Select New Hampshire Locations

LOCATION	TOTALITY START (EDT)
CLARKSVILLE	3:28:12PM
COLUMBIA	3:28:13PM
ERROL	3:29:01PM
LANCASTER	3:29:12PM
NORTHUMBERLAND	3:30:33PM
STRATFORD	3:28:32PM

Eclipse Photography

Photographing an eclipse is an exciting challenge, as the moon's shadow moves near 1,600MPH. There is an element of danger and the pressure of time. Looking at the unfiltered sun through a camera can permanently damage your vision and your camera. If you are unsure, just enjoy the eclipse with specially designed eclipse glasses. Keep a solar filter on your lens during the eclipse and remove for the duration of totality!

Partial Vs. Total Solar Eclipse

To successfully and safely photograph a partial and total eclipse, it is important to understand the difference between the two. A solar eclipse occurs when the moon is positioned between the sun and Earth. The region where the shadow of the moon falls upon Earth's surface is where a solar eclipse is visible.

The moon's shadow has two parts—the penumbral shadow and the umbral shadow. The penumbral shadow is the moon's outer shadow where partial solar eclipses can be observed. Total solar eclipses can only be seen within the umbral shadow, the moon's inner shadow.

You cannot say you've seen a total eclipse when all you saw was a partial solar eclipse. It is like saying you've watched a concert, but in reality, you only listened outside the arena. In both cases, you have missed the drama and the action.

Photographing A Partial And Total Solar Eclipse

Aside from the region where the outer shadow of the moon is cast, a partial solar eclipse is also visible before a total solar eclipse within

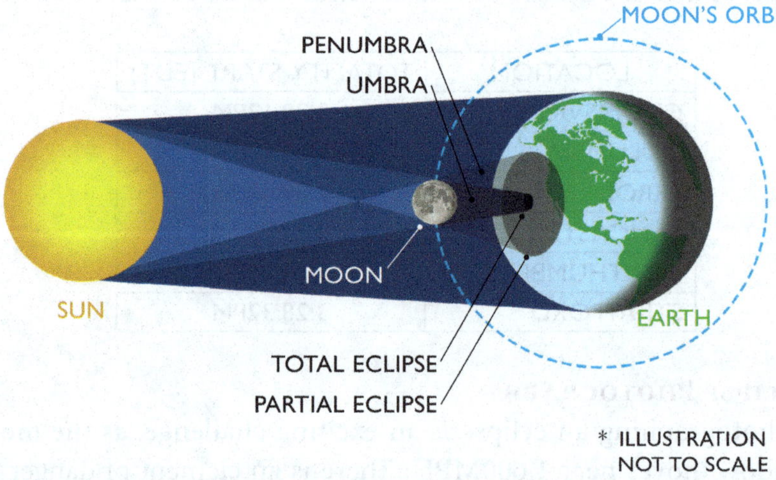

the inner shadow region. In both cases, it is imperative to use a solar filter on the lens for both photography and safety reasons. This is the only difference between taking a partial eclipse and a total eclipse photograph of the sun.

To photograph a total solar eclipse, you must be within the Path of Totality, the surface of the Earth within the moon's umbral shadow.

The Challenge

A total solar eclipse only lasts for a couple of minutes. It is brief, but the scenario it brings is unforgettable. Seeing the radiant sun slowly being covered by darkness gives the spectator a high level of anticipation and indescribable excitement. Once the moon completely covers the sun's radiance, the corona is finally visible. In the darkness, the sun's corona shines, capturing the crowd's full attention. Watching this phenomenon is a breathtaking experience.

Amidst all the noise, cheering, and excitement, you have no more than a few minutes to take a perfect photograph. The key to this is planning. You need to plan, practice, and perfect what you will do when the big moment arrives because there is no replay. The pressure is enormous. You only have a short time to capture the totality and the sun's corona using different exposures.

Plan, Practice, Perfect

It is important to practice photographing before the actual phenomenon arrives. Test your chosen imaging setup for flaws. Rehearse over and over until your body remembers what you will do from the moment you arrive at your chosen spot to the moment you pack up and leave the area.

You will discover potential problems regarding vibrations and focus that you can address immediately. This minimizes the variables that might affect your photographs at the most critical moment.

It's common for experienced eclipse chasers to lose track of what they plan to do. Write down what you expect to do. Practice it time and again. Play annoying, distracting music while you practice. Try photographing in the worst weather possible. Do anything you can to practice under pressure. Eclipse day is not the time to practice.

Once the sun is completely covered, don't just take photographs. Capture the experience and the image of the total solar eclipse in your mind as well. Set up cameras around you to record not just the total solar eclipse but also the excitement and reaction of the crowd.

Eclipse Photography Gear

What do you need to photograph the total eclipse? There are only a few pieces of equipment that you'll need. Preparing to photograph an eclipse successfully takes time. Not only do you have to be skilled and have the right gear, you have to be in the correct place.

Basic Eclipse Photography Equipment

- Solar viewing glasses (verify authenticity)
- Lens solar filter
- Minimum 300mm lens
- Stable tripod that can be tilted to 60° vertical
- High-resolution DSLR
- Spare batteries for everything
- Secondary camera to photograph people, the horizon, etc.
- Remote cable or wireless release

Additional Items

- Video camera
- Video camera tripod
- Quality pair of binoculars
- Solar filters for each binocular lens
- Photo editing software

Equipment to Prepare Before the Big Day

A. Solar viewing glasses

You need a pair of solar viewing glasses as the eclipse approaches.

B. Solar Filter

Partial and total eclipse photography is different from normal photography. Even if only 1% of the sun's surface is visible, it is still approximately 10,000 times brighter than the moon. Before totality, use a solar filter on your lens. Do not look at the sun with your eyes. It can cause irreparable damage to your retinas.

DO NOT leave your camera pointed at the sun without a solar filter attached. The sun will melt the inside of your camera. Think of a magnifying glass used to torch ants and multiply that by one hundred.

C. Lens

To capture the corona's majesty, you need to use a telescope or a telephoto lens. The best focal length, which will give you a large image of the sun's disk, is 400mm and above. You don't want to waste all your efforts by bringing home a small dot where the black disk and majestic corona are supposed to be.

D. Tripod

Bring a stable enough tripod to support your camera properly to avoid unsteady shots and repeated adjustments. Either will ruin your photos. It also needs to be portable in case you need to change locations for a better shot. ***Shut off camera stabilization on a tripod!***

E. Camera

You need to remember to set your camera to its highest resolution to capture all the details. Set your camera to:

- 14-bit RAW is ideal, otherwise
- JPG, Fine compression, Maximum resolution

Bracket your exposures. Shoot at various shutter speeds to capture different brightnesses in the corona. Note that stopping your lens all the way down may not result in the sharpest images.

Choose the lowest possible ISO for the best quality while maintaining a high shutter speed to prevent blurred shots. Set your camera to manual. Do not use AUTO ISO. Your camera will be fooled.

The night before, test the focus position of your lens using a bright star or the moon.

Constantly double-check your focus. Be paranoid about this. You can deal with a grainy picture. No amount of Photoshop will fix a blurry, out-of-focus picture.

F. Batteries

Remember to bring fresh batteries! Make sure that you have enough power to capture the most important moments. Swap in fresh batteries thirty minutes before totality.

G. Remote release

Use a wired or wireless remote release to fire the camera's shutter. This will reduce the amount of camera vibration.

H. Video Camera

Run a video camera of yourself. Capture all the things you say and do during the totality. You'll be amazed at your reaction.

I. Photo editing software

You will need quality photo editing software to process your eclipse images. Adobe Lightroom and Photoshop are excellent programs to extract the most out of your images. Become well versed in how to use them at least a month before the eclipse.

J. Smartphone applications

The following smartphone applications will aid in your photography planning: Wunderground, Skyview, Photographer's Ephemeris, Sunrise and Sunset Calculator, SunCalc, and Sun Surveyor among others.

Camera Phones

Smartphone cameras are useful for many things but not eclipse photography. An iPhone 6 camera has a 63° horizontal field of view and is 3264 pixels across. If you attempt to photograph the eclipse, the sun will be a measly 30-40 pixels wide depending on the phone. Digital pinch zoom won't help here. If you want *National Geographic* images, you'll need a serious camera and lens, far beyond any smartphone.

Consider instead using a smartphone to run a time-lapse of the entire event. The sun will be minuscule when shot on a smartphone. Think of something else exciting and interesting do to with it. Purchase a Gorilla Pod, inexpensive tripod, or selfie stick and mount the smartphone somewhere unique.

Also, partial and total eclipse light is strange and ethereal. Consider using that light to take unique pictures of things and people. It's rare and you may have something no one else does.

Focal Length & the Size of Sun

The size of the sun in a photo depends on the lens focal length. A 300mm lens is the recommended minimum on a full-frame (FF) DSLR. Lenses up to this size are relatively inexpensive. For more magnification, use an APS-C (crop) size sensor. Cameras with these sensors provide an advantage by capturing a larger sun.

For the same focal length, an APS-C sensor will provide a greater apparent magnification of any object. As a consequence, a shorter, less expensive lens can be used to capture the same size sun.

The below figure shows the size of the sun on a camera sensor at various focal lengths. As can be seen with the 200mm lens, the sun is quite small. On a full-frame camera at 200mm, the sun will be 371 pixels wide on a Nikon D810, a 36-megapixel body. A lower resolution FF camera will result in an even smaller sun.

Printing a 24-inch image shot on a Nikon D810 with a 200mm lens at a standard 300 pixels per inch results in a small sun. On this size paper, the sun will be a miserly 1.25 inches wide!

Photographing the eclipse with a lens shorter than 300mm will leave you with little to work with. Using a 400mm lens and printing a 24-inch print will result in a 2.5-inch-wide sun. For as massive as the sun is, it is a challenge to take a large photograph of the sun. The sun will appear to move fast with a 500mm lens, too. Plan to adjust.

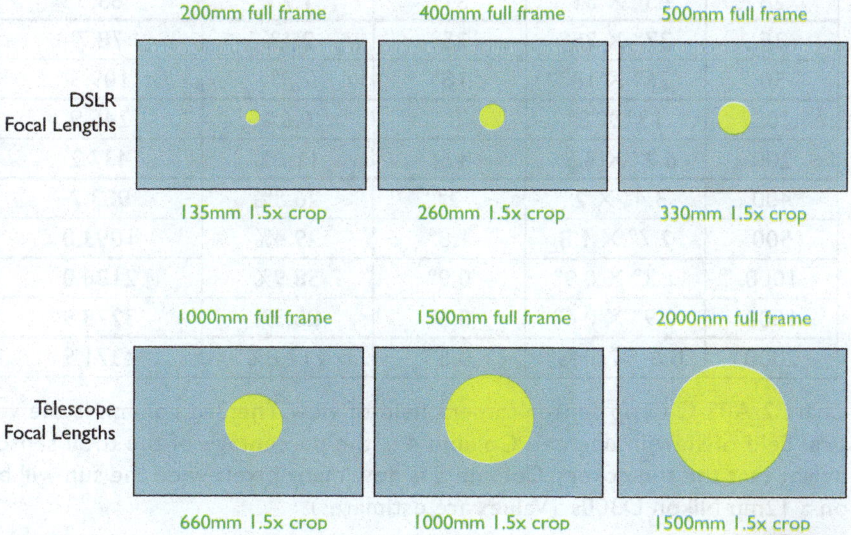

FOCAL LENGTH	FOV FULL FRAME	FF VERT. ANGLE	% OF FF	SUN PIXEL SIZE
14	104° X 81°	81°	0.7%	32.1
20	84° X 62°	62°	0.9%	41.9
28	65° X 46°	46°	1.2%	56.5
35	54° X 38°	38°	1.4%	68.5
50	40° X 27°	27°	2.0%	96.4
105	19° X 13°	13°	4.1%	200.2
200	10° X 7°	7°	7.6%	371.9
400	5° X 3.4°	3.4°	15.6%	765.6
500	4° X 2.7°	2.7°	19.6%	964.2
1000	2° X 1.3°	1.3°	40.8%	2002.5
1500	1.4° X 0.9°	0.9°	58.9%	2892.6
2000	1° X 0.68°	0.68°	77.9%	3828.4

Chart 1: Full-frame camera field of view. The 3rd column is the vertical field of view in degrees. Column 4 is the percentage of the total sensor height that the sun covers. Column 5 is how many pixels wide the sun will be on a 36MP Nikon D810. (Values are estimates)

FOCAL LENGTH	FOV CROP	CROP VERT DEG	% OF CROP	SUN PIXEL SIZE
14	80° X 58°	58°	0.9%	33.9
20	61° X 43°	43°	1.2%	45.8
28	45° X 31°	31°	1.7%	63.5
35	37° X 25°	25°	2.1%	78.7
50	26° X 18°	18°	2.9%	109.3
105	13° X 8°	8°	6.6%	245.9
200	6.7° X 4.5°	4.5°	11.8%	437.2
400	3.4° X 2°	2°	26.5%	983.7
500	2.7° X 1.8	1.8°	29.4%	1093.0
1000	1.3° X 0.9°	0.9°	58.9%	2186.0
1500	0.9° X 0.6°	0.6°	88.3%	3278.9
2000	0.6° X 0.45°	0.5°	117.8%	4371.9

Chart 2: APS-C Crop sensor camera field of view. The 3rd column is the vertical field of view in degrees. Column 4 is the percentage of the total sensor height that the sun covers. Column 5 is how many pixels wide the sun will be on a 12mp Nikon D300s. (Values are estimates)

The big challenge is the cost of the lens. Lenses longer than 300mm are expensive. They also require heavier tripods and specialized tripod heads. The 70-300mm lenses from Nikon, Canon, Tamron, and others are relatively affordable options. It is worth spending time at a local camera shop to try different lenses. Long focal-length lenses are a significant investment, especially for a single event.

To achieve a large eclipse image, you will need a long focal-length lens, ideally at least 400mm. A standard 70-300mm lens set to 300mm will show a small sun. At 500mm, the sun image becomes larger and covers more of the sensor area. The corona will take up a significant portion of the frame. By 1000mm, the corona will exceed the capture area on a full-frame sensor. See the picture in this chapter for sun size simulations for different focal lengths.

Suggested Exposures

To photograph the partial eclipse, the camera must have a solar filter attached. If not, the intense light from the sun may damage (fry) the inside of your camera. This has happened to the author. The exposure depends on the density (darkness) of the solar filter used.

As a starting point, set the camera to ISO 100, f/8, and with the solar filter on, try an exposure of 1/2000. Make adjustments based on the filter used, histogram, and highlight warning.

Turn on the highlight warning in your camera. This feature is commonly called "blinkies." This warning will help you detect if the image is overexposed or not.

Once the Baily's Beads, prominences, and corona become visible, there will only be a few minutes to take bracketed shots. It will take at least eleven shots to capture the various areas of the sun's corona and stars. The brightness varies considerably. No commercially available camera can capture the incredible dynamic range of the different portions of the delicate corona. This requires taking multiple photographs and digitally combining them afterward.

During totality, try these exposure times at ISO 100 and f/8: 1/4000, 1/2000, 1/1000, 1/250, 1/60, 1/30, 1/15, 1/4, 1/2, 1 sec, and 4 sec.

Disable camera/lens stabilization on a tripod!

Photography Time

Set the camera to full-stop adjustments. It will reduce the time spent fiddling. As an example, the author tried the above shot sequence, adjusting the shutter speed as fast as possible.

It took thirty-three seconds to shoot the above 11 shots using 1/3-stop increments. This was without adjusting composition, focus, or anything else but the shutter speed. When the camera was set to full stop increments, it only took twenty-two seconds to step through the same shutter speed sequence. Use a remote release to reduce camera shake.

Assuming the totality lasts less than two minutes, only four shot sequences could be made using 1/3-stop increments. Yet six shot sequences could be made when the camera was set to full stop steps. Zero time was spent looking at the back LCD to analyze highlights and the histogram.

Now add in the bare minimum time to check the highlight warning. It took sixty-three seconds to shoot and check each image using full stops. And that was without changing the composition to allow for sun movement, bumping the tripod, etc. Looking at the LCD ("chimping") consumed **half** of the totality time.

This test was done in the comfort of home under no pressure. In real world conditions, it may be possible to successfully shoot only one sequence. If you plan to capture the entire dynamic range of the totality, you must practice the sequence until you have it down cold. If you normally fumble with your camera, do not underestimate the difficulty, frustration, and stress of total eclipse photography.

Most importantly, trying to shoot this sequence allowed for zero time to simply look at the totality to enjoy the spectacle.

Avoid Last Minute Purchases

You should purchase whatever you think you'll need to photograph the eclipse early. This event will be nothing short of massive. Remember the hot toy of the year? Multiply that frenzy by a thousand. Everyone will want to try to capture their own photo.

Do not wait until the last few weeks before the eclipse to purchase cameras, lenses, filters, tripods, viewing glasses, and associated material. Consider that the totality of the eclipse will streak across

America. Everyone who wants to photograph the eclipse will order at the same time. If you wait until too late to buy what you need, it's conceivable that solar filters to create a total eclipse photo will be sold out in the United States. All filters sold out during the 2017 total eclipse. Whether this happens or not, do not wait to make your purchases. It may be too late.

Practice

You will need to practice with your equipment. Things may go wrong that you don't anticipate. If you've never photographed a partial or total eclipse, taking quality shots is more difficult than you think. Practice shooting the sequence with a midday sun. This will tell you if you have your exposures and timing correct. Figure out what you need well in advance.

Practice photographing the full moon and stars at night. Capture the moon in full daylight to learn how your camera reacts. Astrophotography is challenging and requires practice.

The August 21, 2017, eclipse as seen in Jackson, WY, shot with a Nikon D800 with an 80-400mm lens set to 340mm. The sun is 644 pixels wide on the 7360x4912 image.

This image is shown straight out of the camera without modification. Even with a high-quality camera and lens, photographing an eclipse is challenging.

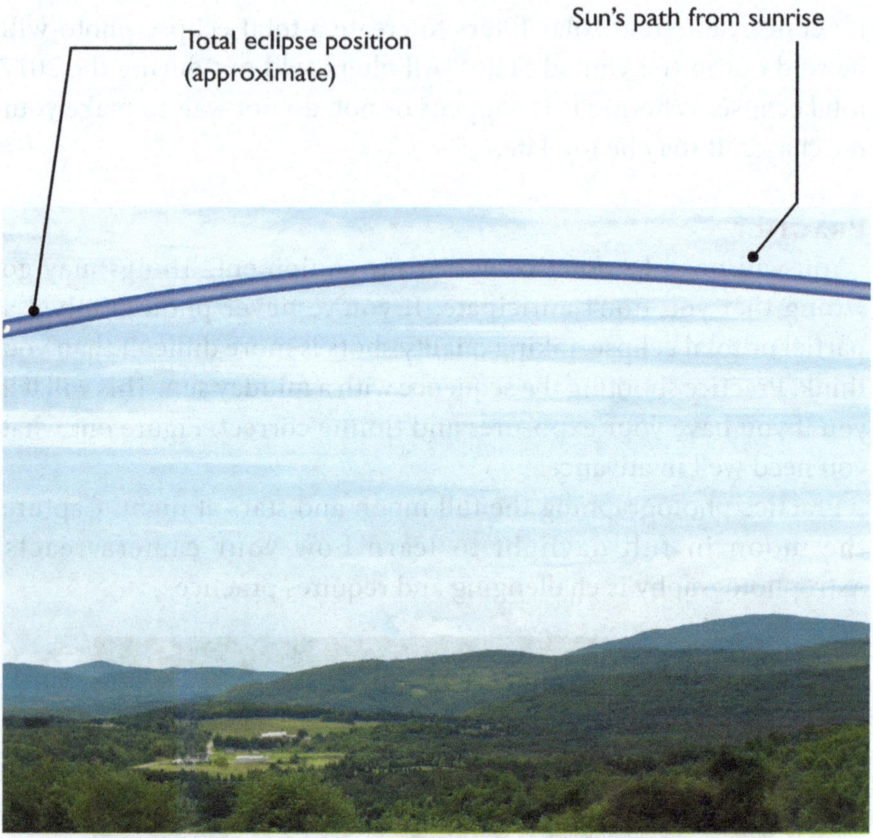

The sun will follow this path on the morning of the eclipse on April 8, 2024. Image of forested area south of Pittsburg.

Note that this image is a simulation and approximation the sun's path and where the total eclipse may appear from one perspective. Refer to the eclipse position data for a more accurate location.

⊙ is the symbol for the sun and first appeared in Europe during the Renaissance.

☾ is the ancient symbol for the moon.

Viewing Locations Around New Hampshire

Tens of thousands of people will travel to and around New Hampshire to view the total eclipse. There are few obstructions and there is a vast amount of space to view the total eclipse from.

If the weather is questionable, seek out a new location as soon as possible. If you wait until the hour before the eclipse, you may find yourself stuck in traffic, as others will be looking for a viewing location. Be safe on the roadways, as drivers may be distracted.

This section contains popular, alternative, and little-known locations to watch the eclipse. As long as there are no clouds or smoke from fires, the partial eclipse will be viewable from anywhere in the state.

Suggested Total Eclipse View Points

Towns and Cities

- Clarksville
- Colebrook
- Columbia
- Dummer
- Errol
- Lancaster
- Northumberland
- Pittsburg
- Stark
- Stewartstown
- Stratford

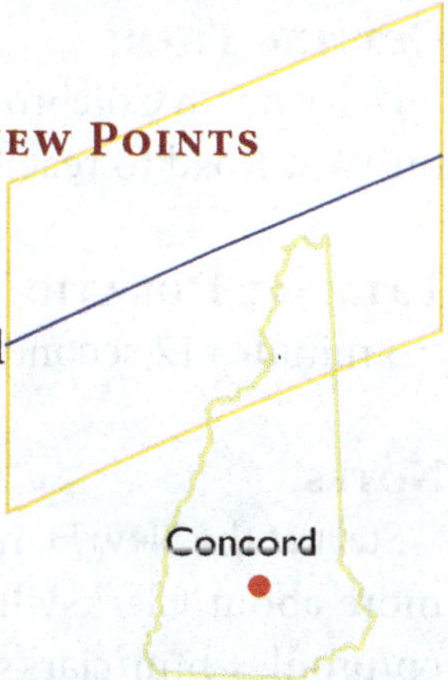

New Hampshire Total Eclipse Path

Unique Locations

- Back Lake
- Beaver Brook Falls Wayside
- Cape Horn State Forest
- Connecticut Lakes State Forest
- Kauffmann Forest
- Lake Francis State Park
- Nash Stream Forest
- Rhubarb Pond
- Umbagog Lake State Park
- Washburn Family Forest

Clarksville

Elevation:	1,981 ft. (604m)
Population:	406
Main road/hwy:	US 3

Overview

Clarksville is the perfect jumping off point for a visit to the Washburn Family Forest, just north of the town along US 3. Originally incorporated in 1853, this small town was actually called the Dartmouth College Grant until 1872. The state legislature deeded the college a substantial land tract in the late 18th century. This small hamlet is the second most northerly town in all of New Hampshire, located only a few miles southwest of Pittsburg.

Getting There

Drive north Colebrook for eleven miles on US 3, then continue east on West Road to reach Clarksville.

Totality Duration

3 minutes 12 seconds

Notes

Start at the New Hampshire Employment Security Website to learn more about Clarksville at https://www.nhes.nh.gov/elmi/products/cp/profiles-htm/clarksville.htm.

Event	Time (EDT)	Altitude	Azimuth
Sunrise	6:12:00AM	0°	78°
Eclipse Start	2:16:56PM	48°	214°
Totality Start	3:28:12PM	39°	235°
Totality End	3:31:24PM	38°	236°
Eclipse End	4:38:38PM	28°	251°
Sunset	7:23:00PM	0°	282°

Colebrook

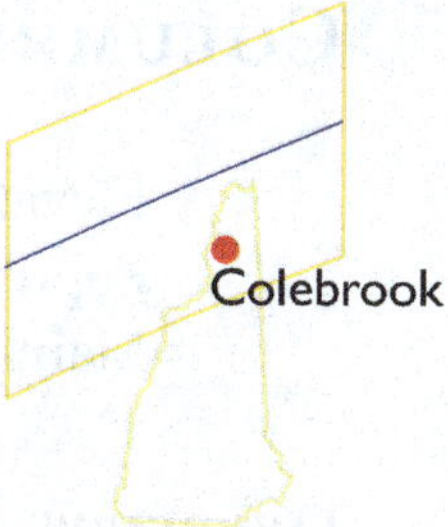

Elevation:	1,020 ft. (311m)
Population:	2,301
Main road/hwy:	US 3

Overview

For those who want to visit Colebrook, there are a number of attractions to entice visitors to this city. The Revolutionary War forge site is here, along with the Colebrook River, which is both a lake and a dam. Hike at the Beaver Brook Falls Wayside and bring a picnic to set up at one of the shelters. There is also the Colebrook Store and the Colebrook Historical Society. The local fair takes place every summer in this city. Enjoy a selection of restaurants and indulge in casual or fine dining after a long day.

Getting There

Drive north from Lancaster on US 3 for thirty-six miles to reach Colebrook.

Totality Duration

2 minute 59 seconds

Notes

Learn more about the historic town of Colebrook at https://www.chamberofthenorthcountry.com/.

Event	Time (EDT)	Altitude	Azimuth
Sunrise	6:12:00AM	0°	78°
Eclipse Start	2:16:48PM	48°	214°
Totality Start	3:28:14PM	39°	235°
Totality End	3:31:14PM	38°	236°
Eclipse End	4:38:37PM	28°	251°
Sunset	7:23:00PM	0°	281°

Columbia

Elevation:	1,007 ft. (307m)
Population:	757
Main road/hwy:	US 3

Overview

The Great North Woods Region is a perfect tourist attraction spot for all who visit this town. You'll love the towering forests and beautiful lakes and rivers. There are so many reasons why tourists visit this area. The main ones are moose watching, snowmobiling, skiing, and even hunting in season. Interestingly, the town claims that many of the local businesses are "home based." There are several family farms and even a few Christmas tree plantations.

Getting There

Columbia is four miles south of Colebrook on US 3.

Totality Duration

2 minutes 55 seconds

Notes

Start your exploration of Columbia at this website: http://www.columbianh.org/.

Event	Time (EDT)	Altitude	Azimuth
Sunrise	6:13:00AM	0°	78°
Eclipse Start	2:16:43PM	48°	214°
Totality Start	3:28:13PM	39°	235°
Totality End	3:31:08PM	38°	236°
Eclipse End	4:38:35PM	28°	251°
Sunset	7:23:00PM	0°	282°

Dummer

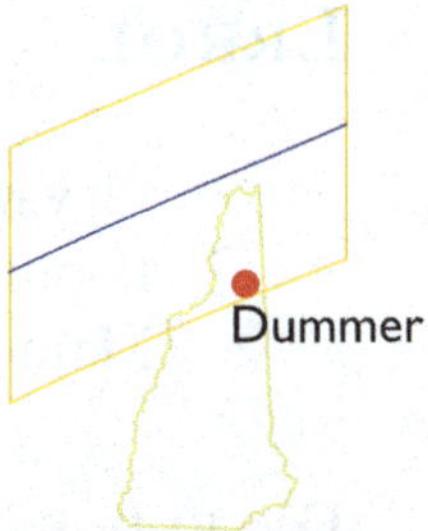

Elevation:	1,135 ft. (346m)
Population:	288
Main road/hwy:	NH 110/16

Overview

Dummer is the perfect jumping off point to watch the total eclipse over the Pontook Reservoir. For those who are visiting this rural town for the first time, there are a few options for lodging during the total eclipse. Dummer has some interesting hotels with scenic views. One of them is the Deer Mountain Lodge and Resort, which is a good year-round getaway in the White Mountains. The Berlin Trails Inn and Farm (*outside of the total eclipse path*) also provides the best scenes for those who want to enjoy a countryside view.

Getting There

Drive east from Northumberland on NH 110/110A/110B to reach Milan, then continue across the river to E of River Rd. and drive three miles north to reach Dummer.

Totality Duration

1 minute 1 second

Notes

Dummer's website is the perfect starting point for your total eclipse venture into northern New Hampshire at http://www.dummernh.com/. *Note that Milan is outside of the total eclipse path.*

Event	Time (EDT)	Altitude	Azimuth
Sunrise	6:12:00AM	0°	78°
Eclipse Start	2:17:03PM	48°	214°
Totality Start	3:29:31PM	38°	236°
Totality End	3:30:32PM	38°	236°
Eclipse End	4:38:54PM	27°	252°
Sunset	7:21:00PM	0°	282°

Errol

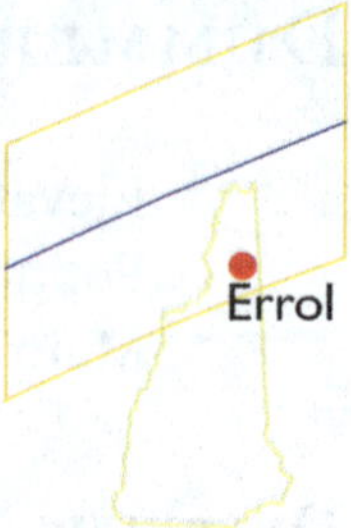

Elevation:	1,236 ft (377m)
Population:	233
Main road/hwy:	NH 16

Overview

From tasty meals to a tour of Cairn O'Mohr Fruit Wines, there are a number of places you can visit in this small town. Head over to Umbagog Lake State Park for wildlife watching, canoeing, fishing, and swimming. The park provides thirty-three remote campsites as well as four remote cabins. All these offer access to the Umbagog Lake National Wildlife Refuge that is located in the Cambridge area of New Hampshire. The cabins and the remote sites provide a variety of wildlife viewing opportunities that include the bald eagles, loons, moose, and river otters. There is also a public campground and a boat launch ride here. Check campground dates well in advance for closures.

Getting There

Drive east from Northumberland on NH 110, then continue north on NH 16 for a total of thirty-five miles to reach Errol.

Totality Duration

2 minutes 12 seconds

Notes

Learn more about Errol at https://www.americantowns.com/coos-county-nh/.

Event	Time (EDT)	Altitude	Azimuth
Sunrise	6:11:00AM	0°	78°
Eclipse Start	2:17:14PM	47°	214°
Totality Start	3:29:01PM	38°	236°
Totality End	3:31:14PM	38°	236°
Eclipse End	4:38:55PM	27°	252°
Sunset	7:21:00PM	0°	281°

Lancaster

Elevation:	862 ft. (263m)
Population:	3,317
Main road/hwy:	US 2/US 3

Overview

This rural town has a darling Main Street for strolling and browsing. Stop in the Weeks Memorial Library, then head over to the Heritage Trail for a walk in the forest where you'll be delighted by an active beaver pond. The Heritage Trail covers over two-hundred thirty miles, part of which travels through Lancaster. It wanders over paved roads, gravel roads, over a covered bridge, and on one side of Mount Prospect. One historical covered bridge of note is the Mechanic Street Bridge, originally built during the American Civil war in 1862. It is on the National Historic Register of Places.

Getting There

Drive north from Concord on I-93 and continue on US 3 for a total of one-hundred two miles to reach Lancaster.

Totality Duration

46 seconds

Notes

Start your exploration of Lancaster at its website at https://www.lancasternh.org/heritage-trail.html.

Event	Time (EDT)	Altitude	Azimuth
Sunrise	6:13:00AM	0°	78°
Eclipse Start	2:16:28PM	48°	214°
Totality Start	3:29:12PM	39°	235°
Totality End	3:29:59PM	39°	236°
Eclipse End	4:38:38PM	28°	251°
Sunset	7:23:00PM	0°	281°

Northumberland

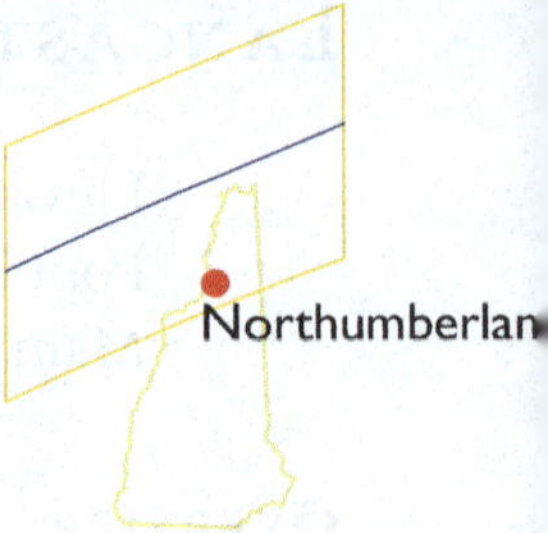

Elevation:	866 ft. (264m)
Population:	2,145
Main road/hwy:	US 3

Overview

You can hike to the top of the ridge on from Cape Horn to view the total eclipse. At 2,060 feet above sea level, this will be a unique location to experience darkness in the middle of the day. There are narrow, steep trails to get to the summit. These are not for the beginner. Avoid the east side as there are no trails or facilities. Learn more about the hike at http://www.franklinsites.com/hikephotos/NewHampshire/capehorn.php.

Getting There

Drive north from Lancaster on US 3 for ten miles to reach Northumberland.

Totality Duration

1 minute 45 seconds

Notes

Explore Northumberland by starting at its website at https://northumberlandnh.weebly.com/.

Event	Time (EDT)	Altitude	Azimuth
Sunrise	6:13:00AM	0°	78°
Eclipse Start	2:16:36PM	48°	214°
Totality Start	3:28:48PM	39°	235°
Totality End	3:30:33PM	38°	236°
Eclipse End	4:38:40PM	28°	251°
Sunset	7:22:00PM	0°	282°

Pittsburg

Elevation:	1,312 ft. (400m)
Population:	890
Main road/hwy:	US 3

Overview

Pittsburg has an abundance of activities both outdoors and indoors. Head outside to Lake Francis State Park for canoeing and fishing. Back Lake's hexagenia hatch is popular with the locals along with trout and bass fishing. Unwind at Tall Timber Lodge or nearby cabins, or stay indoors on a rainy day and browse the general store and other specialty shops. Be on the lookout for Moose in Moose Alley and bird-watching hot spots for loons and eagles.

Getting There

Drive north on US 3 from Lancaster for fifty-four miles to reach Pittsburg.

Totality Duration

3 minutes 15 seconds

Notes

Visit Pittsburg's website at http://www.pittsburg-nh.com/ for updated information on total eclipse events.

Event	Time (EDT)	Altitude	Azimuth
Sunrise	v	0°	78°
Eclipse Start	2:17:02PM	47°	214°
Totality Start	3:28:15PM	38°	235°
Totality End	3:31:30PM	38°	236°
Eclipse End	4:38:41PM	27°	251°
Sunset	7:22:00PM	0°	281°

Stark

Elevation:	987 ft. (301m)
Population:	514
Main road/hwy:	NH 110

Overview

This small town is big on recreation. Visitors find plenty of opportunities for fishing, boating, and hiking. Nearby is the Nash Stream Forest Natural Area where you can hike up the Sugarloaf Mountain trail or enjoy the Jericho Mountain State Park. Situated along Route 110, the town of Stark is known for its historic, beautiful covered bridge and Camp Stark, the World War II POW camp.

Getting There

Drive north on US 3 from Lancaster for nine miles, then continue northeast on NH 110E for another seven miles to reach Stark after turning north and driving over the Stark Covered Bridge.

Totality Duration

1 minute 36 seconds

Notes

Learn more about Stark at their website at http://www.townofstark.com/.

Event	Time (EDT)	Altitude	Azimuth
Sunrise	6:13:00AM	0°	78°
Eclipse Start	2:16:44PM	48°	214°
Totality Start	3:28:59PM	39°	235°
Totality End	3:30:34PM	38°	236°
Eclipse End	4:38:44PM	27°	251°
Sunset	7:22:00PM	0°	281°

Stewartstown

Elevation:	1,174 ft. (358m)
Population:	799
Main road/hwy:	US 3

Overview

Incorporated in 1795, after the Revolutionary War, Stewartstown is one of the many small towns in Coos County. Located along the Connecticut River, the town is a great starting point for trying your had at eclipse photography over a river. If you want to step away from large, crowded cities for the eclipse, consider Coleman State Park in Stewartstown. Bring your camping gear for peaceful fishing, boating, and ATV trails.

Getting There

Drive north Lancaster on US 3 for forty-five miles to reach Stewartstown.

Totality Duration

3 minutes 14 seconds

Notes

Investigate Beecher Falls on the Connecticut River, on the border of New Hampshire and Vermont, at this website http://www.northeastwaterfalls.com/waterfall.php?num=3&p=0.

Event	Time (EDT)	Altitude	Azimuth
Sunrise	6:12:00AM	0°	78°
Eclipse Start	2:16:51PM	48°	214°
Totality Start	3:28:07PM	39°	235°
Totality End	3:31:22PM	38°	236°
Eclipse End	4:38:36PM	27°	251°
Sunset	7:23:00PM	0°	281°

Stratford

Elevation: 862 ft. (263m)
Population: 854
Main road/hwy: US 3

Overview

Stratford is an outdoor paradise. Deer, bear, moose, and other wildlife make Stratford their home. Go hiking along the town's miles of trails and enjoy the wilderness. Fish and canoe in the Connecticut River as you enjoy your time before or after the total eclipse. Stratford is where you will find a Frog Museum at the Foolish Frog. There is an abundance of wildlife around the town. You will have the chance to see wild turkeys, moose, and countless birds.

Getting There

Drive north on US 3 from Lancaster for fifteen miles to reach Stratford.

Totality Duration

2 minutes 11 seconds

Notes

Begin your exploration of activities in and around Stratford at http://www.stratfordnh.org/things-to-do.html.

Event	Time (EDT)	Altitude	Azimuth
Sunrise	6:13:00AM	0°	78°
Eclipse Start	2:16:35PM	48°	214°
Totality Start	3:28:32PM	39°	235°
Totality End	3:30:44PM	38°	236°
Eclipse End	4:38:37PM	28°	251°
Sunset	7:23:00PM	0°	282°

Back Lake

Elevation:	1,575 ft. (480m)
Main road/hwy:	US 3

Overview

If you want to combine an outdoor adventure with your total eclipse experience, Back Lake is an excellent option. Although famous for its trout, you will have to wait until the end of April to begin fishing for trout. Check the updated New Hampshire fishing regulations. Timber Lodge offers boat rentals for you to explore this three-hundred fifty-acre lake. The lodge has multiple buildings, private cottages, and an overall New Hampshire rustic feel to it while still providing modern amenities. It also has all-inclusive dining plans so you can enjoy the eclipse without worrying about bringing too much with you.

Getting There

Drive north on US 3 from Lancaster for fifty-four miles to reach Pittsburg, then continue on for another four miles to reach Back Lake.

Totality Duration

3 minutes 16 seconds

Notes

Visit the Tall Timber Lodge website for updated lodging and travel information at https://talltimber.com/.

Event	Time (EDT)	Altitude	Azimuth
Sunrise	6:12:00AM	0°	78°
Eclipse Start	2:17:06PM	47°	214°
Totality Start	3:28:17PM	38°	235°
Totality End	3:31:33PM	38°	236°
Eclipse End	4:38:42PM	27°	251°
Sunset	7:22:00PM	0°	282°

Beaver Brook Falls Wayside

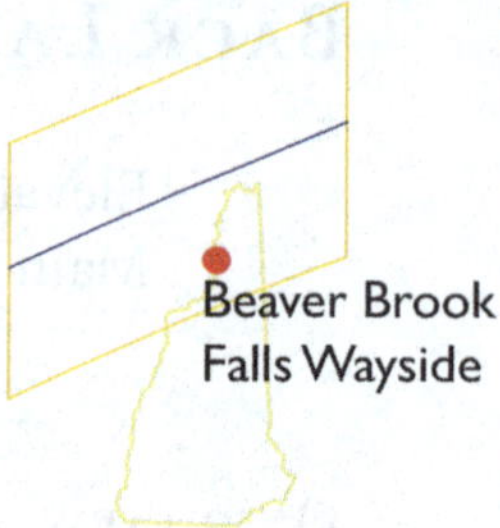

Elevation:	1,158 ft. (353m)
Main road/hwy:	US 3 (via Colebrook)

Overview

The falls at Beaver Brook Falls Wayside will be an enjoyable distraction before and after the total eclipse. The park features a veil-type waterfall cascading over a dark rock face. The walk to the falls is short and family-friendly. It can be enjoyed from the grassy area while sitting on the park bench. The are several picnic tables and a small picnic shelter, along with a group area as well. Make sure to secure any required reservations well in advance of the total eclipse.

Getting There

Drive three miles northeast from Colebrook on Park Street to reach Beaver Brook Falls Wayside.

Totality Duration

3 minutes 2 seconds

Notes

Explore your adventure options for this park at the New Hampshire Parks website at https://www.nhstateparks.org/visit/state-parks/beaver-brook-falls-wayside.

Event	Time (EDT)	Altitude	Azimuth
Sunrise	6:12:00AM	0°	78°
Eclipse Start	2:16:52PM	47°	214°
Totality Start	3:28:15PM	39°	235°
Totality End	3:31:18PM	38°	236°
Eclipse End	4:38:39PM	27°	251°
Sunset	7:23:00PM	0°	282°

Cape Horn State Forest

Elevation:	2,040 ft. (621m)
Main road/hwy:	US 3

Overview

Cape Horn is a narrow, and distinct ridge that is located on the east side of the Pilot Range in New Hampshire. According to some sources, the original name comes from the shape of the ridge line that resembles Cape Horn in South America. It is possible to reach the top of the ridge to view the total eclipse, though it will require some adventurous bushwhacking. Start at either Page Hill Road or Lost Nation Road to reach the edge of the state forest property. From there, make your way toward the ridge line. Do bring navigation and back-country survival and safety gear, as the weather may be unpredictable.

Getting There

Start in Groveton and follow Lost Nation Road until you find a suitable and safe parking spot to begin your backcountry adventure.

Totality Duration

1 minute 28 seconds

Notes

Note that at the time of this writing, the summit is on private property. Secure permission before entering and be prepared for unpredictable outdoor travel.

Event	Time (EDT)	Altitude	Azimuth
Sunrise	6:13:00AM	0°	78°
Eclipse Start	2:16:34PM	48°	214°
Totality Start	3:28:55PM	39°	235°
Totality End	3:30:23PM	38°	236°
Eclipse End	4:38:39PM	28°	251°
Sunset	7:22:00PM	0°	281°

Connecticut Lakes State Forest

Elevation: Varies
Main road/hwy: US 3

Overview

Adjacent to the U.S.-Canadian Border, the Connecticut Lakes State Forest is a rather unique tract of land. It follows US 3, also known as the Daniel Webster Highway, from the Canadian border to the First Connecticut Lake area. There are camping options at Deer Mountain Campground near Second Connecticut Lake where you can enjoy Moose Falls. You can pull out at numerous locations along the road to set up a position to view the total eclipse. Stop by Idlewilde at Second Connecticut Lake with its dirt road loop as another option.

Getting There

Drive north along US 3 from Pittsburg and continue to the Canadian Border to reach this state forest.

Totality Duration

Varies from 3 minutes 25 seconds at the Canadian Border to 3 minutes 16 seconds at First Connecticut Lake.

Notes

Visit the Connecticut Lakes Headwaters Working Forest Website at https://www.nhstateparks.org/visit/state-parks/connecticut-lakes-headwaters-working-forest.

Event	Time (EDT)	Altitude	Azimuth
Sunrise	6:11:00PM	0°	78°
Eclipse Start	2:17:23PM	47°	214°
Totality Start	3:28:28PM	38°	235°
Totality End	3:31:46PM	38°	236°
Eclipse End	4:38:49PM	27°	251°
Sunset	7:22:00PM	0°	282°

Times are for Idlewilde

Kauffmann Forest

Elevation:	Multiple
Main road/hwy:	US 3 / Percy Road

Overview

The Kauffman Forest is a wild tract of land that is also a working forest. This hardwood region has been managed to produce fine wood and also a wild experience for the adventurous. Christine Lake is northeast of the forest and will make for an interesting backdrop for your total eclipse experience. You will likely see loons, osprey, and bald eagles in the area, too. The trail that leads to the top of Devil's Slide will be a good option for nature lovers to find perch to watch the eclipse from.

Getting There

Begin at Stark and drive north along Percy Road which borders Kauffman State Forest.

Totality Duration

1 minute 41 seconds

Notes

Visit the Kauffman Forest web page at https://forestsociety.org/property/kauffmann-forest for updated total eclipse event information.

Event	Time (EDT)	Altitude	Azimuth
Sunrise	6:12:00AM	0°	78°
Eclipse Start	2:16:46PM	48°	214°
Totality Start	3:28:57PM	39°	235°
Totality End	3:30:38PM	38°	236°
Eclipse End	4:38:44PM	27°	251°
Sunset	7:22:00PM	0°	282°

Lake Francis State Park

Elevation:	1,338 ft. (408m)
Main road/hwy:	US 3

Overview

Lake Francis State Park will allow you to view the total eclipse over Lake Francis. The small park has two spur roads that reach forty-five traditional camping sites. There are an additional nine sites with RV water and electric hook-ups, as well as nine specially designated ATV sites, too. As the campground operates with staff between Memorial to Columbus days, camping is first-come, first-served. Bring cash or a check so you can pay for your outdoor adventure.

Getting There

Drive north from Pittsburg along US 3 to Mountain View Cabins, then turn south on Hill Road, left on Glenwood Drive, left on Philbrook Road, and right on River Road to reach the state park.

Totality Duration

3 minutes 13 seconds

Notes

Start at the Lake Francis State Park web page to learn more about the park at https://www.nhstateparks.org/visit/state-parks/lake-francis-state-park.

Event	Time (EDT)	Altitude	Azimuth
Sunrise	v	0°	78°
Eclipse Start	2:17:10PM	47°	214°
Totality Start	3:28:22PM	38°	235°
Totality End	3:31:35PM	38°	236°
Eclipse End	4:38:45PM	27°	251°
Sunset	7:22:00PM	0°	281°

Nash Stream Forest

Elevation: Multiple
Main road/hwy: US 3

Overview

This remote forested area is roughly centered on Nash Stream which runs along the road that traverses the forest and natural area. The forests in the area are a mix of hardwood and softwood of varying ages, providing visitors with a unique wilderness experience. The area is highly popular with outdoor enthusiasts of all kinds. Snow mobilers, hikers, fishers, snow adventurers, and hikers all use this area and maintain it as well.

Getting There

Drive north Stark along Northside Road, then continue on Nash Stream Road to enter the area. The road continues for seven miles into the forested area.

Totality Duration

1 minute 36 seconds (time for Stark)

Notes

Learn more about the Nash Stream Forest with the document at this website https://www.nh.gov/nhdfl/documents/nashstream_mgt-plan_1995.pdf.

Event	Time (EDT)	Altitude	Azimuth
Sunrise	6:13:00AM	0°	78°
Eclipse Start	2:16:44PM	39°	235°
Totality Start	3:28:59PM	39°	235°
Totality End	3:30:34PM	38°	236°
Eclipse End	4:38:44PM	27°	251°
Sunset	7:22:00PM	0°	281°

Times are for Stark

Rhubarb Pond

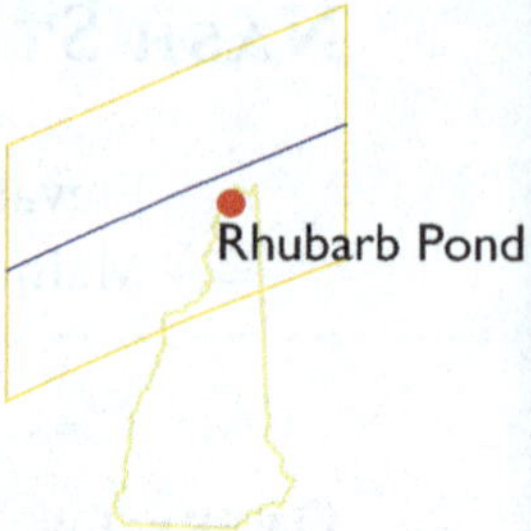

Elevation:	2,303 ft. (702m)
Main road/hwy:	US 3 / Dirt Road

Overview

If you want the northernmost location in New Hampshire to experience the total eclipse, then Rhubarb Pond is where you want to try and be. Located in the remote northern tip of the state, this small fishing pond can be accessed along E. Inlet Road, a lonely dirt track that takes you deep into New Hampshire. Located approximately a mile south of the U.S.-Canadian border, you will be deep inside the woods. There is no official crossing between the two countries. Make sure to bring safety and rescue supplies for you and your vehicle.

Getting There

Drive from US 3 on East Inlet Road (just north of Second Connecticut Lake) for fourteen miles to reach Rhubarb Pond.

Totality Duration

3 minutes 25 seconds

Notes

The road may be inaccessible at the time of the total eclipse. Consult local authorities before venturing to this remote location. https://www.fishingworks.com/new-hampshire/coos-nh/lake/rhubarb-pond/.

Event	Time (EDT)	Altitude	Azimuth
Sunrise	6:10:00AM	0°	78°
Eclipse Start	2:17:35PM	47°	214°
Totality Start	3:28:32PM	38°	235°
Totality End	3:31:57PM	38°	236°
Eclipse End	4:38:52PM	27°	251°
Sunset	7:21:00PM	0°	281°

Umbagog Lake State Park

Elevation:	1,240 ft. (378m)
Main road/hwy:	NH 26

Overview

Umbagog Lake State Park features a multi-use camping area. It has twenty-seven sites available, some with electrical and water hook-ups. There are also thirty-three remote camping sites, three cabins, and four remote cabins accessible by boat only. Make sure to reserve any spots you want well before the total eclipse, as they may be highly-popular with total eclipse chasers attempting to enjoy an adventurous wilderness experience.

Getting There

Drive southeast from Colebrook along NH 26 for twenty-nine miles to reach the state park.

Totality Duration

1 minute 32 seconds

Notes

Learn more about the park's amenities at https://www.nhstateparks.org/visit/state-parks/umbagog-campground.

Event	Time (EDT)	Altitude	Azimuth
Sunrise	6:11:00AM	0°	78°
Eclipse Start	2:17:18PM	47°	215°
Totality Start	3:29:26PM	38°	236°
Totality End	3:30:58PM	38°	236°
Eclipse End	4:39:00PM	27°	252°
Sunset	7:21:00PM	0°	281°

Washburn Family Forest

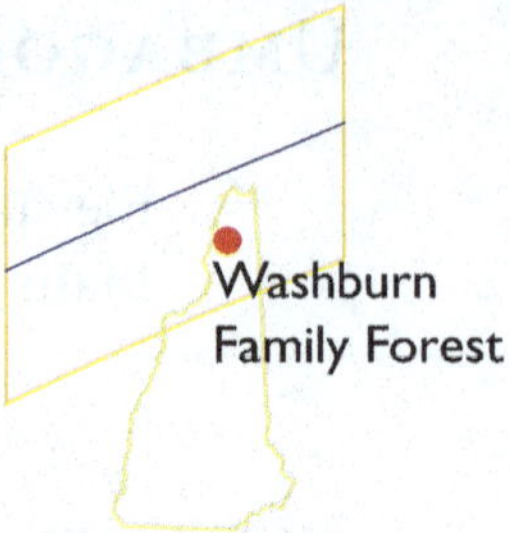

Elevation: Varies
Main road/hwy: US 3

Overview

This two-thousand, one-hundred twenty-eight-acre forested area near Clarksville is a popular destination for snowmobilers and outdoor adventurers. The trailhead to the hiking trail is accessed via a private driveway. This large wilderness area is a habitat for deer, otter, bear, moose, and many species of birds. Fly fishers love the area for the Connecticut River that flows along the northern border of the forest area.

Getting There

Drive southwest from Pittsburg along US 3 for a mile to reach the eastern edge of the forest boundary.

Totality Duration

3 minutes 15 seconds

Notes

Learn more about the forest at the Forest Society's website at https://forestsociety.org/property/washburn-family-forest.

Event	Time (EDT)	Altitude	Azimuth
Sunrise	6:12:00AM	0°	78°
Eclipse Start	2:16:59PM	47°	214°
Totality Start	3:28:12PM	38°	235°
Totality End	3:31:28PM	38°	236°
Eclipse End	4:38:39PM	27°	251°
Sunset	7:23:00PM	0°	281°

REMEMBER THE NEW HAMPSHIRE TOTAL ECLIPSE

April 8, 2024

Who was I with? ____________________

What did I see? ____________________

What did I feel? ____________________

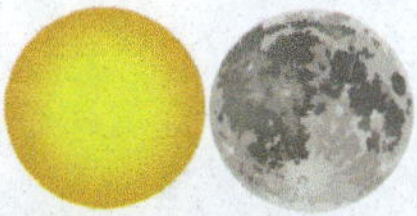

What did the people with me think?

Where did I stay?

Enjoy Other Books by Aaron Linsdau

50 Jackson Hole Photography Hotspots

This guide reveals the best Jackson Hole photography spots. Learn what locals and insiders know to find the most impressive and iconic photography locations in the United States. This is an excellent companion guide to the *Jackson Hole Hiking Guide*.

www.sastrugipress.com/books/50-jackson-hole-photography-hotspots/

Adventure Expedition One
by Aaron Linsdau M.S. & Terry Williams, M.D.

Create, finance, enjoy, and return safely from your first expedition. Learn the techniques explorers use to achieve their goals and have a good time doing it. Acquire the skills, find the equipment, learn to camp, understand medical issues, and learn the planning necessary to pull off an expedition.

www.sastrugipress.com/books/adventure-expedition-one/

Antarctic Tears

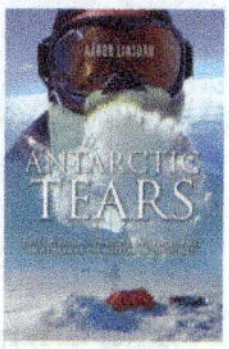

Experience the honest story of solo polar exploration. This inspirational true book will make readers both cheer and cry. Coughing up blood and fighting skin-freezing temperatures were only a few of the perils Aaron Linsdau faced. Travel with him on a world-record expedition to the South Pole.

www.sastrugipress.com/books/antarctic-tears/

How to Keep Your Feet Warm in the Cold

Keep your feet warm in cold conditions on chilly adventures with techniques described in this book. Packed with dozens and dozens of ideas, learn how to avoid having cold feet ever again in your outdoor pursuits.

www.sastrugipress.com/books/how-to-keep-your-feet-warm-in-the-cold/

Jackson Hole Hiking Guide

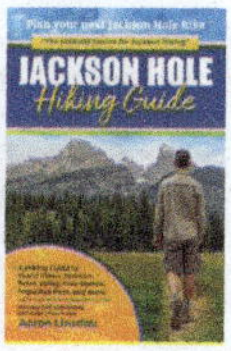

Jackson Hole contains some of the most dramatic and iconic landscapes in the United States. The book shares everything you need to know to hike Jackson's classic trails with canyons, high mountains, and hidden alpine lakes. This book is an excellent companion guide to *50 Jackson Hole Photography Hotspots*.

www.sastrugipress.com/books/jackson-hole-hiking-guide/

Subscribe to Aaron's YouTube channel at www.youtube.com/@alinsdau

If you enjoyed this book, please consider leaving a five-star review and a few words on what you liked about it at your favorite online retailer.

Lost at Windy Corner

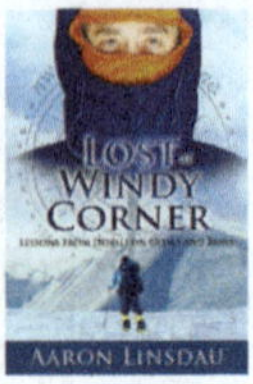

Windy Corner on Denali has claimed fingers, toes, and even lives. What would make someone brave lethal weather, crevasses, and avalanches to attempt to summit North America's highest mountain? Aaron Linsdau shares the experience of climbing Denali alone and how you can apply the lessons to your life.
www.sastrugipress.com/books/lost-windy-corner/

The Motivated Amateur's Guide to Winter Camping

Winter camping is one of the most satisfying ways to experience the wilderness. It is also the most challenging style of overnighting in the outdoors. Learn 100+ tips from a professional polar explorer on how to winter camp safely and be comfortable in the cold.
www.sastrugipress.com/books/the-motivated-amateurs-guide-to-winter-camping/

Two Friends and a Polar Bear
by Terry Williams, M.D. & Aaron Linsdau

This story of friendship is about two old friends who plan to ski across the Greenland Ice Cap along the Arctic Circle in hopes of becoming one of the oldest teams to succeed.
www.sastrugipress.com/books/two-friends-and-a-polar-bear/

About the Author

Aaron Linsdau is the second American to ski alone from the coast of Antarctica to the South Pole (730 miles / 1174 km), setting a world record for surviving the longest expedition ever for that trip. He lead a 310-mile (499 km) ski expedition across the Greenland icecap along the Arctic Circle. Aaron has climbed Denali solo, crossed the Greenland tundra alone, skied across Yellowstone National Park solo, trekked through the Sahara desert, and successfully climbed Mt. Kilimanjaro and Mt. Elbrus in Russia.

Aaron Linsdau at the South Pole.

Use your smart device to scan the QR codes for website links.

Visit www.aaronlinsdau.com/subscribe to learn more about the author. Receive updates when he releases new books and shows.

Visit Sastrugi Press on the web at www.sastrugipress.com to purchase the above titles in bulk. They are available in print, e-book, or audiobook form.

Thank you for choosing Sastrugi Press.

Enjoy Other Books by Sastrugi Press

50 Florida Wildlife Hotspots by Moose Henderson Ph.D.

This is a definitive guide to finding where to photograph wildlife in Florida. Follow the guidance of a professional wildlife photographer as he takes you to some of the best places to see wildlife in the Sunshine State.

www.sastrugipress.com/books/50-florida-wildlife-hotspots/

50 Wildlife Hotspots Grand Teton National Park by Moose Henderson Ph.D.

Find out where to find animals and photograph them in Grand Teton National Park from a professional wildlife photographer. Learn techniques, timing, animal behavior, and composition to create stunning wildlife images.

www.sastrugipress.com/books/50-wildlife-hotspots/

Alaska: A Guide for the Curious by Nikki Mann & Jeff Wohl

Discover the natural world of Alaska. Find out what the plants and animals are like, how to identify them, and what the environment of Alaska is like.

www.sastrugipress.com/books/alaska-a-guide-for-the-curious/

Blood Justice by Tim W. James

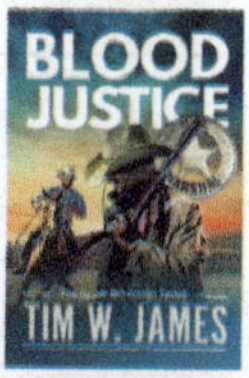

Two brothers, one a preacher's son, the other an adopted would-be slave, set out in opposite directions to avenge their family's murder only to cross paths in pursuit of the killer. Book 1 of the Roger Brinkman Series.

www.sastrugipress.com/iron-spike-press/blood-justice/

Shake Yourself Free by Bob Millsap

Learn how to overcome difficult encounters with misfortune, tragedy, and loss. Emotional recovery is a journey requiring a mindset shift. Get this book now and take control of your life.

www.sastrugipress.com/books/shake-yourself-free/

The Burqa Cave by Dean Petersen

Still haunted by Iraq, a retired soldier seeks solace teaching high school in Wyoming. He soon finds the quiet town is home to murderers, maniacs, and a boy who can see where missing murder victims are. This paranormal thriller-romance surprises readers with unexpected twists.

www.sastrugipress.com/books/the-burqa-cave/

Photography Notes

April 8, 2024

Settings

Successes

Challenges

Photography Notes

April 8, 2024

Camera

Company

Location

Made in United States
North Haven, CT
10 November 2026

22096049R00046